JOURNAL OF
GREEN ENGINEERING

Volume 3, No. 1 (October 2012)

JOURNAL OF GREEN ENGINEERING

Chairperson: Ramjee Prasad, CTIF, Aalborg University, Denmark
Editor-in-Chief: Dina Simunic, University of Zagreb, Croatia

Editorial Board
Luis Kun, Homeland Security, National Defense University, i-College, USA
Dragan Boscovic, Motorola, USA
Panagiotis Demstichas, University of Piraeus, Greece
Afonso Ferreira, CNRS, France
Meir Goldman, Pi-Sheva Technology & Machines Ltd., Israel
Laurent Herault, CEA-LETI, MINATEC, France
Milan Dado, University of Zilina, Slovak Republic
Demetres Kouvatsos, University of Bradford, United Kingdom
Soulla Louca, University of Nicosia, Cyprus
Shingo Ohmori, CTIF-Japan, Japan
Doina Banciu, National Institute for Research and Development in Informatics, Romania
Hrvoje Domitrovic, University of Zagreb, Croatia
Reinhard Pfliegl, Austria Tech-Federal Agency for Technological Measures Ltd., Austria
Fernando Jose da Silva Velez, Universidade da Beira Interior, Portugal
Michel Israel, Medical University, Bulgaria
Sandro Rambaldi, Universita di Bologna, Italy
Debasis Bandyopadhyay, TCS, India

Aims and Scopes
Journal of Green Engineering will publish original, high quality, peer-reviewed research papers and review articles dealing with environmentally safe engineering including their systems. Paper submission is solicited on:

- Theoretical and numerical modeling of environmentally safe electrical engineering devices and systems.
- Simulation of performance of innovative energy supply systems including renewable energy systems, as well as energy harvesting systems.
- Modeling and optimization of human environmentally conscientiousness environment (especially related to electromagnetics and acoustics).
- Modeling and optimization of applications of engineering sciences and technology to medicine and biology.
- Advances in modeling including optimization, product modeling, fault detection and diagnostics, inverse models.
- Advances in software and systems interoperability, validation and calibration techniques. Simulation tools for sustainable environment (especially electromagnetic, and acoustic).
- Experiences on teaching environmentally safe engineering (including applications of engineering sciences and technology to medicine and biology).

All these topics may be addressed from a global scale to a microscopic scale, and for different phases during the life cycle.

JOURNAL OF GREEN ENGINEERING

Volume 3   No. 1   October 2012

M. ČMARADA, R. PETRÁŠ and M. ADAMÍK / General Purpose Self-Managed and Eco-Friendly Wireless Communication Network for Forest Environment   1–11

A.S.G. Andrae / The Effect of Revised Characterization Indices for N_2O and CO_2 in Life Cycle Assessment of Optical Fiber Networks – The case of Ozone Depletion and Aquatic Acidification   13–32

Sara S. MAHMOUD and IMTIAZ AHMAD / Green Performance Indicators for Energy Aware IT Systems: Survey and Assessment   33–69

M. PEJANOVIC-DJURISIC, E. KOCAN and M. ILIC-DELIBASIC / Energy Efficient Wireless Communications through Cooperative Relaying   71–90

TANUJA SATISH DHOPE (SHENDKAR), DINA SIMUNIC and RAMJEE PRASAD / TVWS Radio Spectrum Utilization: Use Case of India-Looking Forward   91–112

Published, sold and distributed by:
River Publishers
P.O. Box 1657
Algade 42
9000 Aalborg
Denmark

Tel.: +45369953197
www.riverpublishers.com

Journal of Green Engineering is published four times a year.
Publication programme, 2012–2013: Volume 3 (4 issues)

ISSN 1904-4720

General Purpose Self-Managed and Eco-Friendly Wireless Communication Network for Forest Environment

M. Čmarada, R. Petráš and M. Adamík

Technical University in Zvolen, 24 T.G. Masaryk Street, 96001 Zvolen, Slovak Republic; e-mail: {michal.cmarada, rudolf.petras, mirysko}@gmail.com

Received: 31 August 2012; Accepted: 25 September 2012

Abstract

Our goal is to create an eco-friendly network for the collection of data from the environment. The network must be energy-independent with the use of renewable energy sources. As a backup power source in the event of unfavourable weather conditions supercapacitors and other environmental friendly energy storage technologies can be used. Our proposed data network needs to be energy efficient and be able to manage itself without any external supervision.

Keywords: communication network, energy harvesting, environment monitoring, eco-friendly technology, energy accumulation, fuel cells.

1 Introduction

Quickly changing climatic conditions and the rapid emergence of modern information and communication technologies provide us with an opportunity to closely monitor the environment around us. By analysing these data we can better understand the environment in which we live, protect it and predict catastrophic events (fires, floods, storms, landslides) or monitor the wildlife.

In this paper, we provide a proposal of self-managed low-power wireless communication network which can be used to transfer different kinds of digital information obtained from various types of sensors and devices in the forest environment. Our goal is that the proposed network is environmentally friendly. We want to achieve this challenge by the exclusion of typical batteries, which contain heavy metals dangerous to the environment. Instead of batteries alternative energy sources can be used and they must be combined in such a way that the unfavourable weather conditions would not cause power failure. This also leads to the need for ecological energy storage, minimization of energy losses and making network operations more efficient as a whole.

2 Network Structure Specification

The communication network itself is one of the most important portions of the system. The main requirements for the communication network are: the use of low-power communication devices, the independence on the central communication network element, energy-efficient data transfer and the ability of the network to function in harsh environment.

Because the network is needed to be operational in the event of failure of several communication modules, it is necessary to design it with respect to the possibility of using alternative communication routes to prevent communication failures (Figure 1 – alternative routes are displayed in different types of lines). Such request complies with the use of unlimited topology also known as "mesh topology". This topology requires a communication module, not only to be able to process their own data, but also the ability to forward data of other elements.

Another important feature of the network is its self-organization. The network must be able to adaptively respond to the change of its internal structure. In case of adding or removing (due to failure) a communication module from the network, it must respond immediately and create alternative communication routes to maintain the communication in the network.

The proposed network shall not depend on a central network element, but at the same time it must allow the co-existence of elements with different configurations (different sensors, transmitters, power sources, etc.). It also needs to use special algorithms for self-organization and self-manageability, which eliminates the need for service interventions in the field and contributes to the long life of the network without any external supervision.

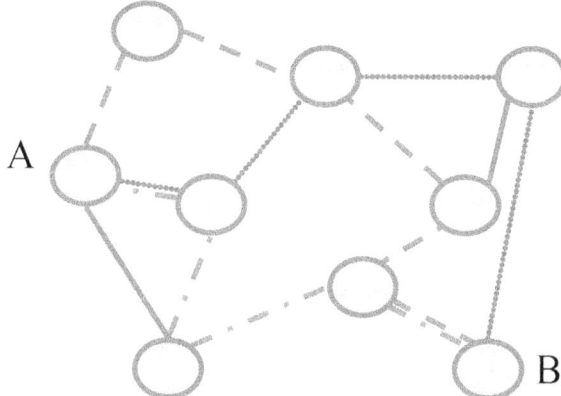

Figure 1 Example of mesh topology with alternate routes from point A to point B.

When creating such algorithms it is necessary to take into account the energy requirements of the network communication especially in case of using energy harvesting systems to collect the energy from the surrounding environment. We have therefore decided to optimize the network in the terms of energy consumption. This ensures an even distribution of energy consumption across the network, where all communication elements will be evenly loaded. Another way to reduce energy consumption of the network is the use of effective compression of transmitted data and compressed sensing, resulting in shortening the time needed to acquire and transfer the data and thus leading to reduction in power consumption.

3 Wireless Communication Technologies

The current massive development of wireless communication technologies is a response to unceasing demand for their use in practice. There are several standards that meet the requirements of our network. They are mostly low-power radio networks such as ZigBee and Bluetooth as well as other systems such as Wi-Fi, GSM, etc.

Despite the continuous trend of increasing transmission speed, these communication technologies with low data rates and low power consumption represent huge and successful development in forestry.

It is advised to use ZigBee communication standard in these networks, which provides the transmission speeds up to 250 kbps and a range around 100 meter in the basic version and up to 1500 meter in Pro version. High reliability and long battery life (several years) makes it suitable for creating this type of network in the forest environment.

4 Network Sensor Module

Proposed communication network basically consists of a distributed system of intelligent sensor modules with elements of versatility and adaptability. The main task of the module is the collection of information from sensors in the environment and sending them through the network to a remote computational centre. Sensor modules should cover a wide range of functions, which are often used in forestry. The internal structure of this module is shown in block diagram (Figure 2), and includes information processing and energy management parts.

The sensor module offers processing power of microcontroller (MCU) which manages scanning data from sensors, processing of these data and sending them in a suitable form for further processing by high-frequency radio module. An important task is the organization and optimization of network traffic for example using fuzzy control and power management in cooperation with the energy management circuit. This circuit provides energy harvesting and also the storage and distribution of the accumulated energy. Recently, technologies have emerged, which support wireless transfer of small quantities of energy. This can serve to create a certain kind of energy channel for wireless power transmission.

4.1 Network Management

Advanced method is the use of fuzzy control (Figure 3) in decision-making processes and determining optimal communication paths and power management. This type of control uses humanlike logic and is suitable for use in situations where the output parameter is influenced by several input parameters. As input parameters in this case, we can use for example path length, path load and energy consumption of communicational devices. The output in this case will then be the parameter determining the route metrics. Using this parameter it is then possible to establish the optimum route chart.

4.2 Energy Harvesting

Energy harvesting (recuperation) from the environment is a relatively new scientific discipline that deals with the collection of small quantities of en-

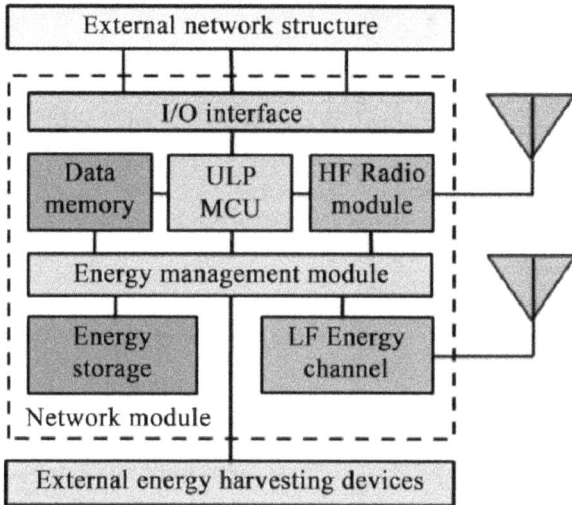

Figure 2 Block diagram of network sensor module.

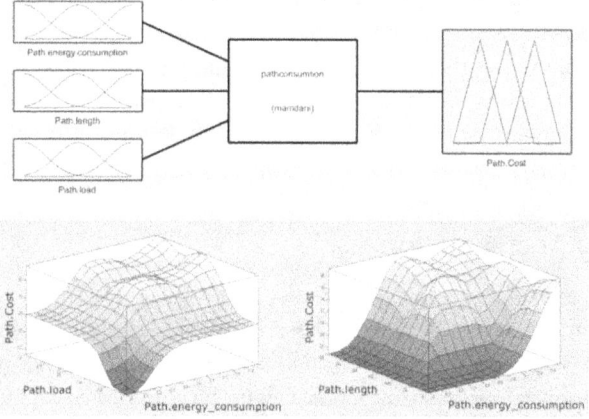

Figure 3 Example of fuzzy control for determining path metric.

ergy dissipated in the surrounding environment. In the domestic, work or forest environment many different kinds of energy can be found in many different forms such as light, heat, vibration, fluid flow, etc. [1]. Similarly the EM fields, which the environment is currently overloaded, can serve such a

Table 1 Alternative energy sources [1].

Energy sources	Description of sources	Efficiency	Power gain
Mechanical vibrations industry/organisms	Frequency 1 to 1 kHz Acceleration 1 to 10 m·s^{-2}		100 μW/cm^2
Photovoltaic energy	Exterior 100 mW/cm^2	10 to 24%	10 mW/cm^2
	Interior 0.1 mW/cm^2	10 to 24%	10 μW/cm^2
Thermal energy	Industry 100 mW/cm^2	3%	10 mW/cm^2
	Organism 20 mW/cm^2	0.1%	25 μW/cm^2
EM field 0.9 GHz	0.3 μW/cm^2	50%	0.1 μW/cm^2
GSM 1.8 GHz	0.1 μW/cm^2		
EM field, Wi-Fi	10 μW/cm^2	50%	0.01 μW/cm^2

purpose. Different types of energy sources with their energy gain are shown in Table 1.

These energy sources can be converted into electricity using micro-generators of electric energy. These consist of physical-electrical converters, which generate low levels of voltage. To power CMOS electronic circuitry it is necessary to increase the voltage levels for them to operate correctly. Management of power supply and distribution for application circuitry is covered by power management circuit.

An example is the thermoelectric micro-generator which uses the Peltier module for generating voltage from the temperature difference and specialized power management circuit as converter of voltage levels, which serves for the collection, storage and distribution of electricity.

4.3 Energy Accumulation

In terms of alternative environmentally friendly energy storage, there are several systems that can be used. These systems currently represent modern and advanced technology, which is also reflected in their price, yet they have found their application in various fields, which require the use of clean energy sources. One of the most publicized alternative energy sources is hydrogen, which in combination with fuel cells appears to be the energy source of the future. Thanks to hydrogen it possible to store large amounts of energy for a long time without losses.

Another environmentally friendly method of energy storage is the re-placement of batteries for supercapacitors, which are sufficient to power low-power devices and is often used for example in solar systems. Their main

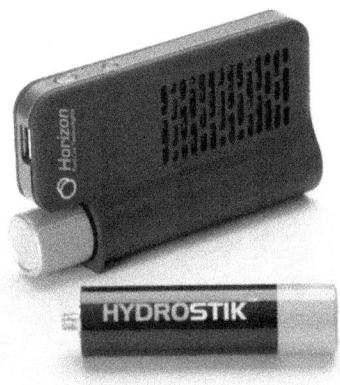

Figure 4 Hydrostik fuel cell [2].

advantage is fast charging and the ability to deliver large amounts of energy in a very short time. The downside is less stored energy per unit of mass.

4.3.1 Accumulation of Energy by Hydrogen

Hydrogen is an ideal element that can be used for energy storage. Due to the existence of fuel cells hydrogen can be used as a power source and due to the existence of devices such as reversible fuel cell or PEM electrolyzers, it is possible to store energy in hydrogen form. For a long time the safe storage of hydrogen for later use was difficult. Currently, there are ways to store hydrogen in stable solid form (such as Hydrostik from Horizon fuel cell technologies) without the danger of explosion. This is a battery-like device (Figure 4), which contains a special metal alloy that allows hydrogen to be stored in a solid-state. This device provides 15 Wh of energy [2].

4.3.2 Accumulation of Energy by Supercapacitors

Unlike ordinary batteries supercapacitors store energy in an electric field and not in a chemical reaction. Through the use of advanced materials and due their inner construction, supercapacitors can store large amounts of energy.

Compared with ordinary batteries they achieve a much lower density of energy conservation. It is possible to use them in applications with low power requirements and systems using energy harvesting from the surrounding environment. Such systems allow their use for short-term energy storage. They

Figure 5 Supercapacitor/Ultracapacitor [3].

are also used in combination with fuel cells in many industrial applications [4].

4.3.3 Independent Energy Channel

Advantageous technology is creating a parallel energy transfer channel to the existing network topology. This channel would allow the transfer of electricity wirelessly between communication modules. An example of wireless transfer of electrical energy can be "WiTricity" (invented by a team of MIT physicists, led by Professor Marin Soljacic). This principle would allow the transport of electricity from modules with a higher energy gain to modules with a smaller energy gain and more power consumption. This system is ideal in environments where independent power supply system is used and allows energy harvesting from the environment [5, 6].

5 Application Possibilities and Advantages of the Network

It is possible to formulate a number of areas in which our concept of wireless eco-friendly sensor networks can be fully implemented:

- *Monitoring of Environmental Quality* – where relevant data (e.g. pollution levels, the presence of hazardous elements, etc.) is mostly collected at the monitoring stations.

- *Monitoring of meteorological parameters* – will help to produce quality information on the meteorological situation in the area. It can help to identify the conditions for the protection of forests and countries, but also in the evaluation of conditions for overpopulation of pests in forests.
- *Monitoring of wildlife migration* – The issue of wildlife migration monitoring is especially important for animal species protection. Based on the knowledge of the territory it is possible to create new protected areas or increase/decrease the existing level of protection.
- *Alert and early warning mechanism* – forest fires, earthquakes, landslides, volcanoes, floods... Dangers of forest fires are currently being monitored in several European countries. These systems are not yet connected to other systems for monitoring extraordinary events. If we implement a system for monitoring of multiple natural disasters, it will give us the possibility of creating a coherent universal protection system in the country against extraordinary events.

6 Benefits and Contributions

The contribution from the project will be in several areas:

- *Theoretical field* – to acquire new knowledge of energy-independent networks with energy generation from renewable energy sources such as solar and wind energy, mechanical vibrations, heat and others.
- *Methodological and educational field* – aims to bring new knowledge about energy-efficient eco-friendly wireless networks.
- *Field of implementation*
  - Outcomes of the project will be publications directly for practice, but also for theoretical, pedagogical and research sector.
  - Design networks models which will use different energy sources for its operation.

7 Conclusion

The proposed self-managed low-power wireless network can be used in the data acquisition system from the environment and transmission of different types of digital data. Application possibilities of this network have the ability to create high added value in quality and quantity of valuable data obtained from multiple devices. The proposed network is able to deliver useful knowledge in predicting and monitoring of extraordinary events, environmental and wildlife monitoring and early warning systems.

Based on the international exchange of experience in this project it is possible to create unified system for evaluation and processing of meteorological data, data from climate changes, early warning systems and others. This synergistic effect of uniform evaluation and processing system is an important added value of our project.

References

[1] J. Šuriansky and M. Hrčková. Mikrozdroje elektrickej energie pre napájanie autonómnych senzorických systémov. In Acta Facultatis Technicae, TU Zvolen, 2010.

[2] Horizon. Fuel cell technologies. Available online: http://www.horizonfuelcell.com/.

[3] Maxwell. BOOSTCAP ultracapacitor product family. Available online: www.maxwell.com/ultracapacitors/.

[4] Maxwell. Hydrogenics and Maxwell combine technologies to bring the material handling equipment market improved productivity with clean, high performance fuel cell power. Case study.

[5] A. Kurs, A. Karalis, R. Moffatt, J.D. Joannopoulos, P. Fisher, and M. Soljačić. Wireless power transfer via strongly coupled magnetic resonances. Science Magazine, 317, 2007.

[6] A. Karalis, J.D. Joannopoulos, and M. Soljačić. Efficient wireless non-radiative mid-range energy transfer. Annals of Physics, 323(1), 2008.

Biographies

Michal Čmarada was born in Liptovský Mikuláš (Slovakia) in 1984. He attended the Secondary Technical School of Jozef Murgaš in Banská Bystrica and later continued his study at the University of Žilina, Faculty of Management Science and Informatics in the field Computer Engineering. He received his M.Sc. degree in 2009. Currently he is a PhD student at Technical University in Zvolen. His research interests include 3D image processing and scanning in manufacturing technology.

Rudolf Petráš was born in Zvolen (Slovakia) in 1985. He attended the Secondary Technical School of Jozef Murgaš in Banská Bystrica. Then He continued his study at the University of Žilina, Faculty of Management Science and Informatics in the field Computer Engineering. Now he is a PhD. student at Technical University in Zvolen.

Miroslav Adamík was born in Zvolen (Slovakia) in 1986. He received his M.Sc. degree in mechatronics from Technical University in Zvolen, Slovakia, in 2010. He is currently working on his Ph.D. thesis on micro-electro-mechanical (MEMS) sensors utilization for manufacturing technology control systems supporting virtual reality environment. His current research interests include industrial automation, wireless sensor networks, energy harvesting technology and computer science as a programming and virtual reality design.

The Effect of Revised Characterization Indices for N_2O and CO_2 in Life Cycle Assessment of Optical Fiber Networks – The Case of Ozone Depletion and Aquatic Acidification

A.S.G. Andrae

Huawei Technologies Sweden AB, Skalholtsgatan 9, 16494 Kista, Sweden;
e-mail: anders.andrae@huawei.com

Received 22 August 2012; Accepted: 24 September 2012

Abstract

The general trend for fixed broadband is that FTTx will overtake ADSL platforms and the number of FTTx subscribers is increasing exponentially. Moreover, it is likely that the ozone depletion potential (ODP) of dinitrogen oxide (N_2O) and the aquatic acidification potential (AAP) of CO_2 have been underestimated in LCA studies. The aim of this study is for the first time to assess the ODP and AAP of different FTTx network deployments in Italy adding the most recent characterization factors for N_2O and CO_2. An LCA case study was conducted covering three FTTx deployments (FTTC is compared to FTTB and FTTH) for 10,000 homes during one year in Italy. The focus is primarily on ODP and AP results in different life cycle phases. The ODP results suggest, using 0.017 kg/kg instead of 0 kg/kg as CFC-11e factor for N_2O, for the greenfield/high power customer premise equipment (CPE) scenario, that FTTB, FTTC, and FTTH all rises from around 80–100 gram to around 600–700 gram CFC-11e/year dominated by the use and deployment stages. For AAP, with 1.752 kg/kg as SO_2e factor for CO_2 instead of 0 kg/kg, the rise is from 5–6 tons to 1,500–1,800 tons SO_2e/year. The weight of the use stage is increasing with these new characterization indices. For FTTC

Journal of Green Engineering, Vol. 3, 13–32.

controlling the power of the CPE is more important than the technique used for deployment. However, for FTTB and FTTH the deployment technique becomes almost as important as the power mode. Concerning FTTH, the main drivers for CFC-11e footprint are the electricity usage of the home gateways (HGWs), their manufacturing, and the use of diesel trucks in traditional civil works and mini-trench deployment. The inclusion of "average" bandwidth gives an advantage for FTTH as more data can be transferred more efficient and faster. For brownfield deployment in Italy (low power CPEs), FTTH architecture has the lowest amount of total CFC-11e emissions (appr. 130 grams). One of the most important criteria, from ozone depletion point of view, when choosing an FTTx network, is whether fiber has been deployed or not. Including the ODP factor for N_2O increases the ODP score by 430–660% for the present systems. The increase for AAP is dramatic and shall be interpreted as a suggestion to include CO_2 acidification in further LCIA research.

Keywords: aquatic acidification, CO_2, FTTx, FTTH, life cycle assessment, N_2O, optical fiber networks, ozone depletion.

Notation

AAP   Aquatic Acidification Potential
BF   Brownfield
FTTx   Fiber To The X
GF   Greenfield
ODP   Ozone Depletion Potential
RMA   Raw Material Acquisition

1 Introduction

Global communication based on ICT is rapidly increasing. The EU project "Energy Aware Radio and Networking Technologies" (EARTH) has predicted that the data traffic between 2010 to 2020 will rise by something like 1,700% and the number of ICT Equipment in use by around 100% [1]. The general trend for fixed broadband is that fiber to the X (FTTx, x = Cabinet, C, Home, H, Building, B) will overtake asymmetric digital subscriber line (ADSL) platforms as the number of FTTH subscribers is increasing exponentially [2]. Another trend is that carbon emissions caps or taxes probably will be sooner or later introduced in a formal way. Moreover, within the ICT sector several standardization efforts have been finalized [3, 4]. Taking this into account, Telecom Italia and Huawei jointly performed a streamlined life cycle assessment (LCA) study in order to estimate the carbon footprint of the introduction of three different FTTx networks [5, 6]. Through the LCA study it was possible to propose an optimization the system under study by finding the energy usage and carbon emission "hot-spots". In life cycle simulations carbon dioxide emissions are rather straightforward to estimate compared to other footprints and the energy used is often fossil based. As the databases and LCA methodologies are improved, the introduction of other footprints will be trivial. As research produce more knowledge our understanding of LCIA indices are revised. Recently Lighthart et al. [7] showed that improved understanding of LCIA indices in the CML baseline 2000 LCIA method (CML) is highly relevant for LCA studies using zinc products. Here the ozone depletion potential (ODP by CML) of the FTTx networks is discussed in light of new facts about the underestimated CFC-11e footprint of N_2O [8]. In the same manner the aquatic acidification potential of CO_2 (AP by CML) is explored.

The problem to be addressed is: What are the implications for an LCA study of FTTx networks of adding characterization indices for N_2O for ODP and CO_2 for AAP in CML?

2 Materials and Methods

The present research is based on a previously performed LCA [1, 5, 6] and an LCA from 2008 by FTTH Council Europe for an FTTH network where the functional unit was "allow a European citizen to use FTTH technologies during one year" [9]. The scope included production of passive equipment such as optical fiber cables and boxes and active equipment such as optical network terminals (ONT), optical line terminals (OLT) deployment of cables, use of network (ONTs and OLTs), and incineration/landfill of cables. The environmental benefits per year were also estimated and thereby the result is displayed as the number of years it would take to "pay" for the environmental loadings caused by the FTTH network. FTTH networks are usually deployed in a number of characteristic topographies. The amount of optical fiber cable and deployment technique will differ according to topography. For example, for "Urban Dense" topography FTTH Council Europe used 60% re-use of existing infrastructure, 20% traditional civil works, and 20% micro-trenching. "Urban Wide" and "Rural Deployment" use different shares of deployment techniques. Anyway, the depreciation for a scenario of 60% Urban Dense,

30% Urban Wide and 10% Rural Deployment was 9.6 years as far as ODP [9, section 7.1.1]. Assuming a service life of 5 years for active equipment and 30 years for the remainder, the FTTH network would generate 0.005 g CFC-11e (ODP) and 90 g SO2e (AAP) per subscriber per year and micro-trench deployment would be the main contributor. In this research for the deployment impacts per type and distance the impact assessment numbers of FTTH Council Europe were used.

The added value of the present paper is the first case study to our know-ledge introducing a "new" LCIA factor for CO_2 for AAP. Also the N_2O (0.017 kg CFC-11e/kg) developed by Ravishankara et al. [8] within ODP has not been used much [10, 11].

The role of CO_2 in *lake/sea* acidification is not as clear as for *ocean* acidification [12, 13]. Eutrophication also enhances the CO_2 acidification [13].

2.1 LCA Basics

LCA is a standardized method [3, 4, 14, 15] for making model based estimations of the environmental exchanges associated with technology functions.

LCA studies are required to follow four main steps:

1. Goal & scope definition,
2. Inventory analysis,
3. Impact assessment of the inventory, and
4. Interpretation of the impact assessment.

The impact assessment is usually done with so called mid-point and/or end-point valuation.

This paper will highlight some advancement in impact assessment com-pared to a previous carbon footprint study [5, 6].

2.2 Goal & Scope

The *studied product system* (SPS) is shown in Figure 1 denoted by the dashed line. The architectural scope would be rather wide if all system nodes of a fixed network would be included and allocated to the specific network. The transport and core network equipment such as local area network (LAN) switches and routers are not parts of the SPS as well as the personal com-puters (PCs) and already installed copper cables. Included is the fixed access network from the OLT on the Central Office to the ONU/ONT on the user

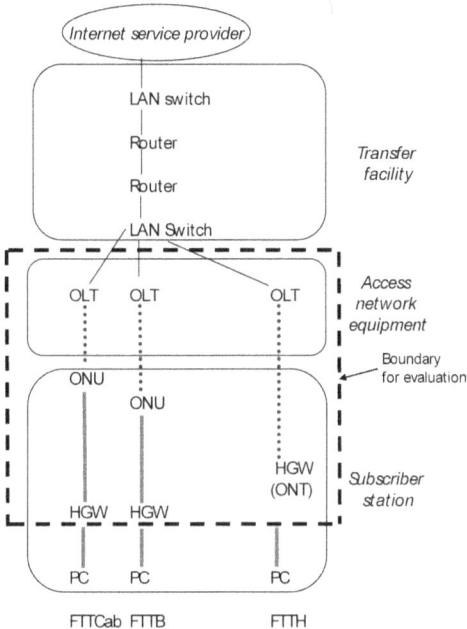

Figure 1 Studied product system of LCA study.

side as well as the optical distribution network (ODN) connecting them. The scope was chosen to highlight differences in between FTTx technologies. The dotted lines between OLT & ONU and OLT & HGW represent optical fiber calbes. The full lines between ONU & PC and HGW & PC, respectively, represent copper cables.

The LCA study performed has neither included the life cycles of internet service provider and the transfer facility, nor the PC. For the environmental LCA the functional unit (f.u.) is *broadband network in an Italian urban dense area for use by 10,000 homes during one year*, and the system boundaries are from cradle-to-grave.

Summary of SPS, cut-off and excluded building blocks:

- SPS: RMA+Production and Use of HGW, ONT, OLT, optical fiber, Deployment of optical fiber, End-of-life treatment of hardware and optical fibers.
- Cut-off from SPS: End-of-life treatment of hardware and optical fibers, cooling of central offices, support equipment RMA+Production+EoLT

Table 1 Assumptions for greenfield and brownfield deployment in Italy.

Deployment type	Network type		
	FTTC	FTTB	FTTH
	Brownfield scenario/Greenfield scenario (%)	Brownfield scenario/Greenfield scenario (%)	Brownfield scenario/Greenfield scenario (%)
Mini-trench	15/35	0/25	0/25
Traditional civil works	15/35	0/25	0/25
Existing deployment	70/30	100/50	100/50

- Excluded: End-user equipment such as PCs, C&C Network, Data centers, Service Provider activities.

Scenario development is unavoidable in LCA studies and here a greenfield and brownfield scenario for deployment was set according to Table 1.

Typical for ICT networks is that the lifetime of system parts vary and this has to be handled when expressing the result annually as required by the recent ETSI and ITU LCA standards [3, 4]. The lifetime of the studied FTTx networks was assumed to be 30 years and therefore the amounts of different hardware and cables is proportional to their lifetime. For example, 26 tons of fiber cables are deployed for FTTH, but per year only 0.87 ton is used as the fiber cables can be in the ground for 30 years. Per year only 2,000 home gateways (HGW (ONT), Figure 3) are used as their lifetime was assumed to be 5 years.

The purifying and drawing of optical fiber in its production were excluded. Other excluded parts are splitters and distribution boxes. These parts were excluded as they likely are a small share of the total score. These omitted parts constitute the so called "cut-off" from the SPS. Anyway, the greenfield scenario is a kind of "worst-case" scenario regarding power usage, running of digging machines and amount of deployed optical fiber cable. The detailed architecture and parameters of FTTx solutions are also defined. FTTH case is shown in Figure 2 as an example.

2.3 Inventory Analysis

Starting from the power usage measurements of three different networks [16], the scope was expanded in order to include manufacturing of hardware, transport of hardware from China to Italy, deployment of each network, and end-of-life. Maintenance & repair during the use phase are excluded.

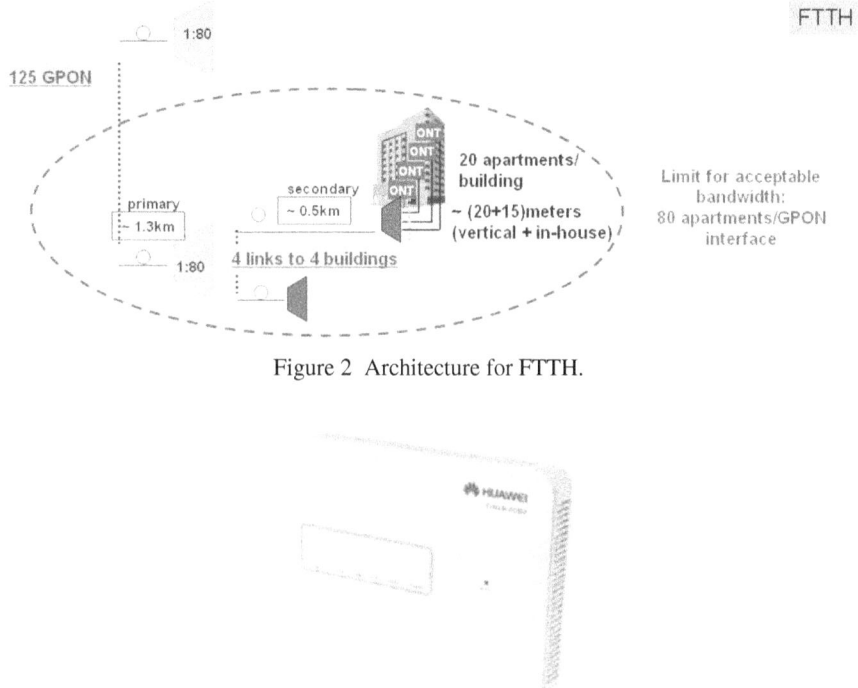

Figure 2 Architecture for FTTH.

Figure 3 One piece of HGW (ONT) (EchoLife HG863).

The assembly of the hardware is assumed to be entirely in China. Huawei provide the OLTs (Huawei product name *MA5600T*), ONUs (*MA5603* and *MA5606*), outdoor cabinets for ONU (*B01D200* and *F01E100*), HGW (ONT) equipment (*EchoLife HG863*, Figure 3) for FTTH as well as HGW very-high-bitrate DSL (VDSL) modems (*EchoLife HG520v*) used by FTTC and FTTB.

The cradle-to-gate analysis of optical fiber cables is likely underestimating the purification and drawing processes. The model is based on Unger and Gough material content [17] and appropriate processes from the LCA tool SimaPro 7.3.2.

Concerning the digging methods in deployment, three different solutions have been considered: mini-trench, traditional civil works and usage of existing channels. For FTTH, FTTB, and FTTC, 412.5, 315 and 65 km optical fiber cable is deployed, respectively. For example the distance deployed cable

Table 2 Summary of life cycle inventory for FTTH networks per functional unit.

Substance	Unit	Indicative Uncertainty	FTTC	FTTB	FTTH
CO_2	kg	±30%	880,000	1,000,000	810,000
CFC-11	kg	±16%	0.01	0.04	0.05
Halon 1211	kg	±88%	0.007	0.008	0.007
Halon 1301	kg	±74%	0.002	0.002	0.002
CH_4	kg	±76%	38	90	48
N_2O	kg	±50%	31	35	28
SO_2	kg	±21%	3,600	4,100	3,400
NH_3	kg	±57%	28	31	21
NO_2	kg	±54%	1	3	14
NO_x	kg	±54%	2,600	2,800	1,700

for FTTH is calculated as:

$$G \times (D_P \times R_P + D_S \times R_S \times L) \tag{1}$$

where G is the number of (gigabit capable passive optical networks) GPONs ($= 125$); R_P is the average reduction coefficient for primary network, between 0.5 (best) and 1 (worst) ($= 1$); R_S is the average reduction coefficient for secondary network ($= 1$); D_P is the primary distance optic fiber [km] ($= 1.3$); D_S is the secondary distance optic fiber [km] ($= 0.5$); L is the number of links ($= 4$) (see Figure 2).

The average reduction coefficient, R, reflects the share of links which are included under the same digging.

In Table 2 a summary of the inventory analysis for Greenfield/High power scenario is shown.

The CO_2 values in Table 2 are lower than the ones in [5, 6] due to lower contribution per km from deployment.

On a high level, the LCA model for FTTH network consists of 14 main modules (Tables 3 and 4). The amount needed of each LCA module is decided by each individual FTTH deployment. Concerning the digging methods in deployment, three different solutions have been considered: mini-trench, traditional civil works and usage of existing ducts.

Perhaps the most interesting observation from Table 3 is the 690% increase for Italian (average retrospective) electricity production. This is of general relevance as electricity is used in most LCA studies.

Table 4 strongly suggests that the role of CO_2 in LCIA calculations which involve aquatic acidification needs to be clarified further.

Table 3 Summary of life cycle CFC-11e emissions of main LCA modules for FTTH network.

Name	Unit	g CFC-11e ($N_2O = 0$) per unit	g CFC-11e ($N_2O = 0.017$ kg/kg) per unit	% increase due to N_2O new ODP index
1. Optical fiber production	km (1km weighs appr. 34 kg)	0.14	0.16	14
2. Transport model from China to Italy	ton×km	0.36	0.64	78
3. Existing infrastructure deployment	km	~0	1	not applicable
4. Mini-trench deployment	km	10	21	110
5. Traditional civil works deployment	km	4	8.3	110
6. Electricity, Italy (average retrospective)	kWh	5.2×10^{-5}	4.1×10^{-4}	690
7. HGW (ONT) production	kg	1.9×10^{-3}	2.7×10^{-2}	1,360
8. OLT production	kg	1.8×10^{-3}	1.4×10^{-2}	680

3 Results

3.1 Impact Assessment

Figures 4 and 5 show some ODP results for the greenfield and brownfield scenarios, respectively. The production includes manufacturing of network equipment and transport from China to Italy. Deployment includes manufacturing of site materials (e.g., concrete) and deployment operations.

For FTTH, when adding the CFC-11e factor for N_2O, the relative weight of Deployment (GF) decreases from 44 to 18% and the ICT Equipment Use increases from 48 to 72%.

For all networks, the production, deployment and use phases are more important than others. The use stage is doubtlessly the main contributor to CFC-11e footprint for the greenfield deployment scenario.

As shown in Figure 5 for the greenfield deployment, the CFC-11e results for calculated by LCA are (on average) 40% higher than results calculated by multiplying the electricity usage by a CFC-11e emission factor.

Table 4 Summary of life cycle SO_2e emissions of main LCA modules for FTTH network.

Name	Unit	kg SO_2e ($CO_2 = 0$) per unit	kg CFC-11e ($CO_2 = 1.752$ kg/kg) per unit	% increase due to new CO_2 AP index
1. Optical fiber production	km (1 km weighs 34 kg)	0.91	213	23,300
2. Transport model from China to Italy	ton	9.7	1,800	18,400
3. Existing infrastructure deployment	km	1.6	346	21,500
4. Mini-trench deployment	km	88	33,200	37,600
5. Traditional civil works deployment	km	62	12,800	20,500
6. Electricity, Italy (retrospective)	kWh	0.003	1.1	36,600
7. HGW (ONT) production	kg	0.96	66	6,800
8. OLT production	kg	0.23	32	13,800

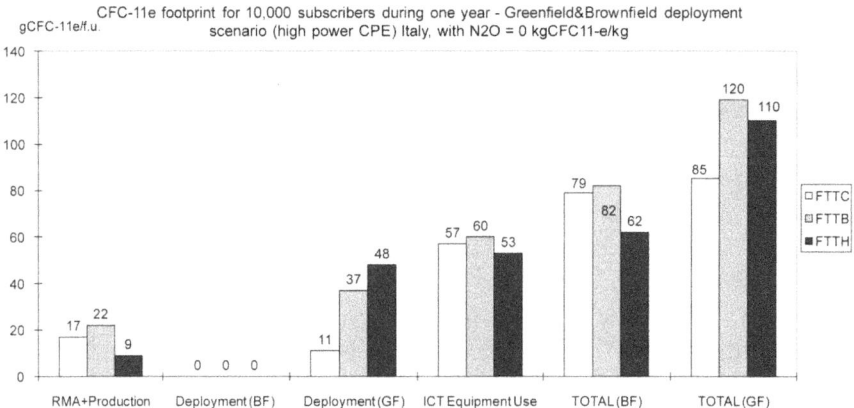

Figure 4 CFC-11e footprint with $N_2O = 0$ CFC11-e kg/kg for the greenfield and brownfield deployment in Italy.

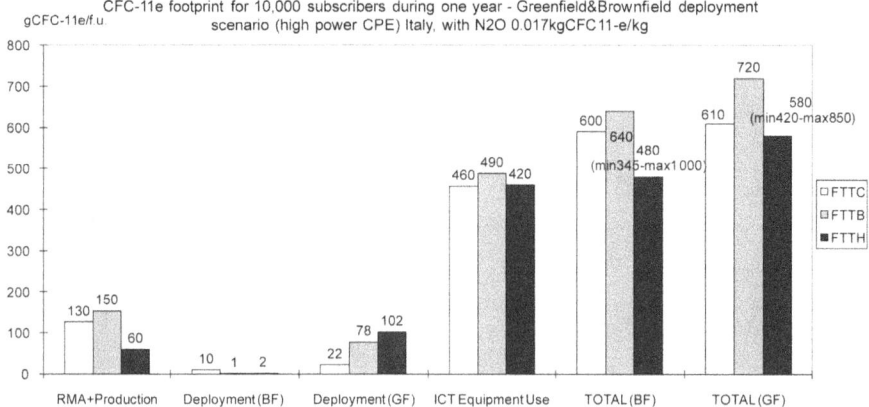

Figure 5 CFC-11e footprint with $N_2O = 0.017$ kg CFC11-e/kg for the greenfield and brownfield deployment in Italy.

3.2 Aquatic Acidification of CO_2

Potentially acidifying emissions are particularly SO_2 and NO_2 which in LCIA are aggregated based on their H^+ formation potential.

$$SO_2 + H_2O \rightarrow 2H^+ + SO_4^2 \qquad (2)$$

$$NO_2 + \frac{1}{2}H_2O \rightarrow H^+ + NO_3^- \qquad (3)$$

SO_2 is commonly the basis for acidification in LCIA and other compounds are expressed in SO_2-equivalents (SO_2e).

Acidification potential of SO_2 = two moles H^+/molecular mass of SO_2 = $2/64.0644 = 0.031$ H^+/g SO_2 which corresponds to 1 g SO_2e/g in LCIA methods.

Acidification potential of NO_2 = 1 mole H^+/molecular mass of NO_2 = $1/46 = 0.021$ H^+/g NO_2. $0.021/0.031 = 0.7$ g SO_2e/g.

$$CO_2 + H_2O \rightarrow 2H^+ + CO_3^{2-} \qquad (4)$$

One mole of CO_2 produces two moles of H^+. Acidification potential of CO_2 = two moles H^+/molecular mass of CO_2 = $2/44 = 0.0454$ H^+/g CO_2. $0.0454/0.031 = 1.46$ g SO_2e/g.

The characterization factor for SO_2 in AAP in CML is 1.2, hence CO_2 could have a factor of 1.752.

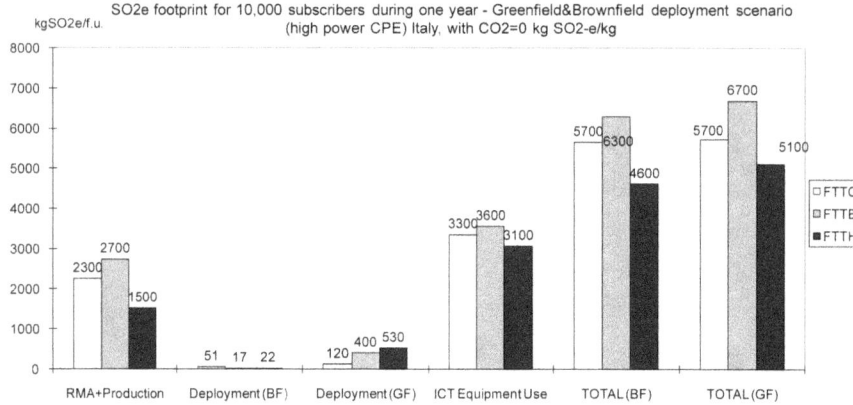

Figure 6 SO$_2$e footprint with CO$_2$ = 0 SO$_2$e/kg for the greenfield and brownfield deployment in Italy.

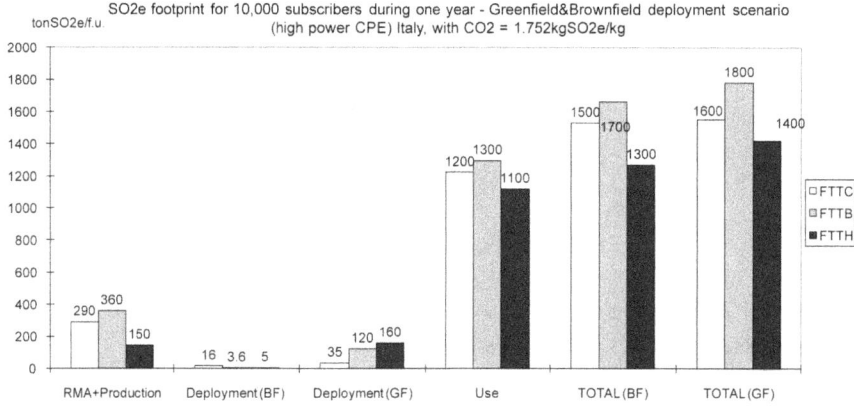

Figure 7 SO$_2$e footprint with CO$_2$ = 1.752 kg SO$_2$e/kg for the greenfield and brownfield deployment in Italy.

From this the simplification is done that CO$_2$ has 1.46 times the aquatic acidification of SO$_2$ in LCIA methods. The implications for acidification in CML baseline 2000 are shown in Figures 6 and 7.

For FTTH, when adding the SO$_2$e factor for CO$_2$, the relative weight of Deployment (GF) is about the same 10%, but the RMA+Production share decreases from 30 to 10% and the ICT Equipment Use increases from 60 to 78%.

4 Discussion

The most important part of an LCA is the interpretation which includes contribution, uncertainty, and sensitivity analyses in which the robustness of the results is tested.

According to contribution analysis (Figures 4–7) the most important phase is Use followed by RMA+Production, and Deployment. Specifically for FTTH the most contributing processes are Italy electricity production and diesel burnt in building machines.

4.1 Interpretation – Uncertainty

The uncertainty analysis in LCA investigates how the precision of used data influence the spread of the final score. The difference between the systems was shown to be enough to draw conclusions as the LCA tool SimaPro quantifies the process correlation. However, the uncertainties of ODP and AAP indices were not included here.

4.2 Interpretation – Sensitivity

To further exemplify the sensitivity analysis, below the greenfield deployment/high power CPE scenario is compared to a more realistic brownfield deployment/low power CPE scenario. For the alternative brownfield deployment assumption, FTTB and FTTH use 100% and FTTC uses 70% of existing infrastructure (Table 1). The realistic low power hypothesis is that the HGWs are on full mode for four hours and in low power mode for 20. This reduced the electricity usage by more than 80% for these HGWs. For ODP LCIA score of FTTH, the brownfield deployment/low power scenario is 77% better than the greenfield deployment/high power CPE scenario. Figure 8 shows the combined effect of deployment technique and power mode for CPEs.

For FTTC controlling the power of the CPEs is more important than the technique used for deployment.

For FTTB and FTTH the deployment technique becomes more important. Moreover, for the brownfield deployment/high power CPE scenario, i.e., when optical fibers have already been deployed, FTTH is better (480 g) than FTTC (600 g) and FTTB (640 g). This means that an important criteria from CFC-11e point of view, when choosing an FTTx network, is whether fiber has been deployed or not.

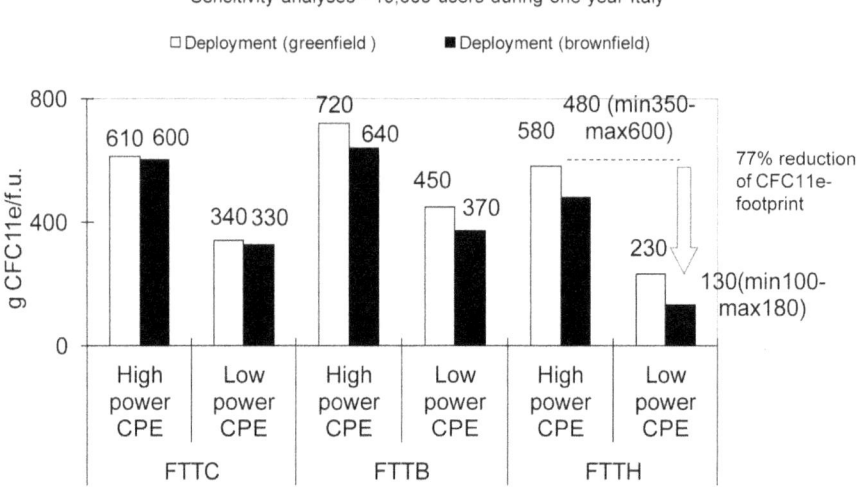

Figure 8 Combined effect of deployment technique and power mode for all networks in Italy.

The brownfield deployment/low power scenario also highlights FTTH as the winning architecture (130 gram compared to 370 for FTTB and 330 for FTTC).

To move into the key message for LCA, Figure 9 shows the effect of using the revised characterization index for N_2O within the ODP category in CML baseline 2000 LCIA method. Furthermore, Figure 9 shows a strong indication of how CO_2 acidification is underestimated in LCIA calculations. The acidification of oceans caused partly by anthropogenic CO_2 emissions to lakes/sea is a strong reason for reductions of global CO_2 [12, 13].

The result for FTTH is of comparable magnitude (per user) as FTTH Council Europe [9]. From Figures 9 and 10 it is clear that 0.011 g CFC-11e and 500 g SO_2e per user per year are the values to be compared to 0.005 and 90 g, respectively, from FTTH Council Europe [9].

A further indication of the effect on end-point methods of including CO_2 index for AAP and N_2O index for ODP is shown in Figure 11.

For the end-point method Eco-Indictator'99 (H) [18] with no modifications of Ozone depletion and Acidification indices they are together around less than 1% of the total scores in Figure 11 (left-hand side).

When adding new proportional values modifying the Eco-Indicator'99, 1.51986 $PDF*m^2yr/kg$, for CO_2 in "Acidification/Eutrophication" and

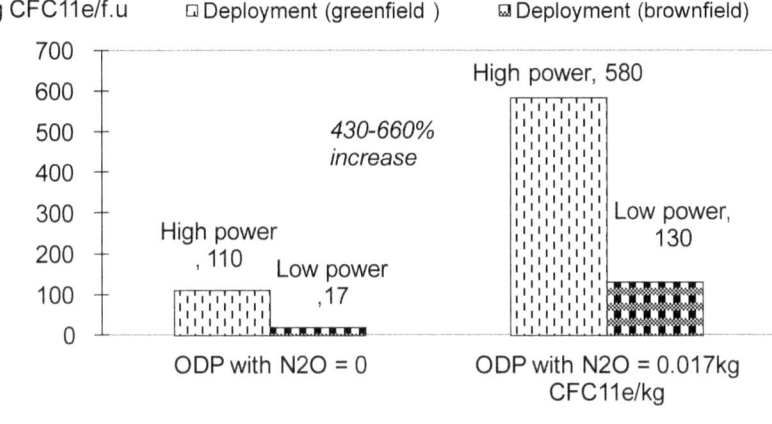

Figure 9 Effect of including N_2O index for FTTH scenarios.

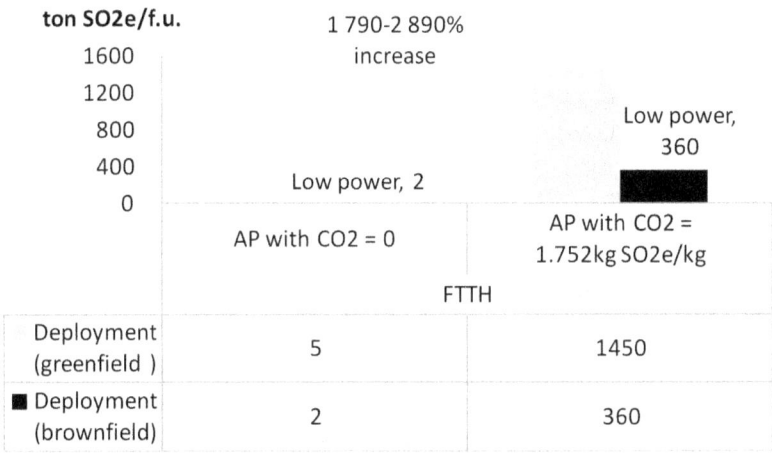

Figure 10 Effect of including CO_2 index for mid-point AAP for FTTH scenarios.

1.785×10^{-5} DALY/kg for N_2O in "Ozone layer", "Acidification/ Eutrophication" increases from 1.1 to 37.5%, whereas "Ozone layer" remains insignificant. This shows that these environmental effects are valued differently in Eco-Indicator'99 (H). The uncertainties of end-point Eco-indicator scores in Figure 11 are not modelled here, however, they are higher than mid-point uncertainties.

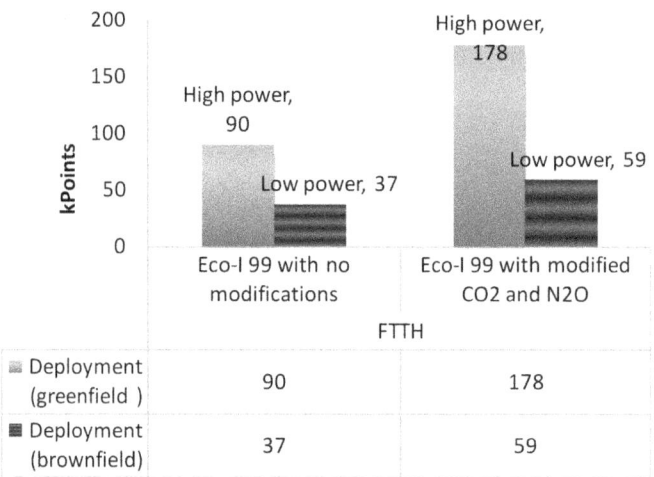

	Eco-I 99 with no modifications	Eco-I 99 with modified CO2 and N2O
Deployment (greenfield)	90	178
Deployment (brownfield)	37	59

Figure 11 Effect of including N_2O and CO_2 indices for end-point method Eco-Indicator'99 (H) for FTTH scenarios.

4.2.1 CFC-11e-Efficiency Analysis

CFC-11e-efficiency is defined here as "average" bandwidth provided by each FTTx network divided by CFC-11e emissions for each annual FTTx network user. As shown in Figure 12, FTTH is considerably more efficient ("more output than input") than especially FTTC. FTTH is by far the best option even when the starting point is the greenfield deployment/high power CPE scenario. Concerning FTTH, the main drivers for CFC-11e footprint are the electricity usage of the HGWs, their manufacturing, and the use of diesel trucks in mini-trench and traditional civil works deployment. The inclusion of "average" bandwidth expressed as megabit per second (Mb/s) gives an advantage for FTTH as more data can be transferred more efficient and faster.

End-of-life processes seem to be irrelevant from a CFC-11e emission point of view and have therefore not been shown explicitly.

Robust LCA studies support two major benefits. The first is Benefit Maximization (most environmental reduction for least economic cost) and the second Continuous Improvement. In the first case the LCA shows where most environmental loadings occur and then the decision makers can find the most cost efficient solution to reduce these environmental loadings. In the second case, the first study helps to clarify the goals for the next product generation.

This analysis has highlighted that it is not enough to base the CFC-11e footprint on electricity usage measurements alone.

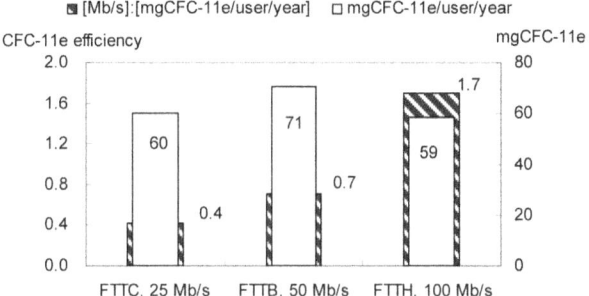

Figure 12 CFC-11e efficiency analysis of FTTx networks in Italy for Greenfield/high power CPE scenario.

Moreover, this study has shown that it is also not enough to compare the life cycles for the hardware systems alone, even though that is a necessary starting point for further research.

5 Conclusions

For the first time FTTC, FTTB and FTTH broadband networks have been simultaneously compared focusing of ODP and AAP by using LCA with a functional unit of *broadband network in an Italian urban dense area for use by 10,000 homes during one year*. The results of this analysis show that, for brownfield deployment in Italy (low power CPEs), FTTH architecture has the lowest amount of total CFC-11e emissions (appr. 130 g). One of the most important criteria, from ozone depletion point of view, when choosing an FTTx network, is whether fiber has been deployed or not. FTTH is more CFC-11e-efficient than FTTC and FTTB. Including the ozone depletion potential factor increases the ODP score by 430–660% for the present systems.

6 Recommendations and Perspectives

In order to get a more comprehensive understanding of the environmental implications of the present networks, water impact categories could be analysed. As the databases and LCA methodologies are improved, the introduction of other footprints (such as water footprint) will be trivial. This will on the other hand demand more primary data collection than CFC-11e footprint estimations. Naturally all LCAs including mid-point ODP analyses should include the "new" index for N_2O [8].

The LCIA field is challenging as new insights about environmental mechanisms are constantly appearing in other research fields than strict LCA.

For global warming likely the current understanding of which are the relevant contributing gases and land usages is correct. Global Warming Potential (GWP) results are therefore rather reliable emphasized by recent integration of the temporal dynamics of global warming was integrated with LCA [19].

Here the two previously *known* effects, N_2O ozone depletion and CO_2 aquatic acidification, are explored and an attempt was done to show how LCA results for ODP and AAP will be affected. The inclusion of N_2O in mid-point ODP is obviously necessary, however the CO_2 influence on AAP scores is probably not as strong as examined here. Likely instead in LCIA a new mid-point impact category for *oceanic acidification potential* is needed.

Still many challenges for LCIA modelling lies ahead of which these are just a few:

1. the indirect ozone depletion potential of water vapour driven by CO_2 and CH_4 emissions [20],
2. the global dimming (mitigating global warming) of particulates [21],
3. the ozone (O_3) creating potential (of black carbon and NO_x) which causes global warming [22], and
4. the influence of the above on environmental damage cost calculations [23].

From a Service LCA perspective, it was beyond the scope of the case study to find out if FTTH is also more effective (accomplished work compared to planned target) in specific working situations than FTTB and FTTC. To investigate such issues LCAs of ICT Services are needed. These can be performed with ETSI [3], ITU [4] and Greenhouse Gas Protocol [24].

Acknowledgement

Support from Huawei Technologies Co. Ltd., Telecom Italia, is gratefully acknowledged.

References

[1] A.S.G. Andrae. European LCA Standardization of ICT: Equipment, networks, and services. In M. Finkbeiner (Ed.), Towards Life Cycle Sustainability Management, 1st edn., pp. 483–493. Springer, Berlin, 2011.
[2] F. Effenberger and T.S. El-Bawab. Passive optical networks (PONs): Past, present, and future. Optical Switching and Networking, 6(3):143–150, 2009.

[3] European Telecommunications Standards Institute. ETSI TS 103 199 V1.1.1 (2011-11) Environmental Engineering (EE); Life Cycle Assessment (LCA) of ICT equipment, networks and services; General methodology and common requirements. 2011. URL: http://www.etsi.org/deliver/etsi_ts/103100_103199/103199/01.01.01_60/ts_103199v010101p.pdf. Accessed: August 20, 2012.

[4] International Telecommunication Union, L.1410 (03/12). Methodology for the assessment of the environmental impact of information and communication technology goods, networks and services. 2012.

[5] G. Griffa, L. Radice, A.S.G. Andrae et al. Carbon footprint of next generation fixed networks. In Proceedings of 32th International Telecommunications Energy Conference, INTELEC 2010, Orlando, FL, June 6–10, Session 15:4, 2010.

[6] Huawei Communicate. Carbon efficiency evaluation of FTTx deployment. Issue 57. September 2010.

[7] T.N. Ligthart, R.H. Jongbloed, and J.E. Tamis. A method for improving Centre for Environmental Studies (CML) characterisation factors for metal (eco)toxicity – The case of zinc gutters and downpipes. International Journal of LCA, 15(8):745–756, 2010.

[8] A.R. Ravishankara, J.S. Daniel, and R.W. Portmann. Nitrous oxide (N_2O): The dominant ozone-depleting substance emitted in the 21st Century. Science, 326(5949):123–125, 2009.

[9] FTTH Council Europe. Developing a generic approach for FTTH solutions using LCA methodology. Methodological guide, Final version. February 2008.

[10] A.S.G. Andrae and O. Andersen. Life cycle assessment of integrated circuit packaging technologies. International Journal of LCA, 16(3):258–267, 2011.

[11] J. Lane and P. Lant. Including N_2O in ozone depletion models for LCA. International Journal of LCA, 17(2):252–257, 2012.

[12] S.C. Doney, V.J. Fabry, R.A. Feely, and J.A. Kleypas. Ocean acidification: The other CO_2 problem. Annual Review of Marine Science, 1:169–192, 2009.

[13] W.G. Sunda and W.J. Cai. Eutrophication induced CO_2-acidification of subsurface coastal waters: Interactive effects of temperature, salinity, and atmospheric PCO_2. Environmental Science and Technology, 46(19):10651–10659, 2012.

[14] International Standardisation Organisation. ISO 14040 – Environmental management – Life cycle assessment – Principles and framework, 2006.

[15] International Standardisation Organisation. ISO 14044 – Environmental management – Life cycle assessment – Requirements and guidelines, 2006.

[16] C. Bianco, F. Cucchietti, and G. Griffa. Energy consumption trends in the next generation access network – A Telco perspective. In Proceedings of 29th International Telecommunications Energy Conference, INTELEC 2007, Rome, Italy, 30 September–4 October, pp. 737–742, 2007.

[17] N. Unger and O. Gough. Life cycle considerations about optic fiber cable and copper cable systems: A case study. Journal of Cleaner Production, 16(4):1517–1525, 2008.

[18] M. Goedkoop, P. Hofstetter, R.M. Wenk, and R. Spriemsma. The ECO-indicator 98 explained. International Journal of LCA, 3(6):352–360, 1998.

[19] A. Kendall. Time-adjusted global warming potentials for LCA and carbon footprints. International Journal of LCA, 17(8):1042–1049, 2012.

[20] J.G. Anderson, D.M. Wilmouth, J.B. Smith, and D.S. Sayres. UV dosage levels in summer: Increased risk of ozone loss from convectively injected water vapor. Science, 337(6096):835–839, 2012.

[21] M. Wild. Global dimming and brightening: A review. Journal of Geophysical Research, 114, D00D16, doi:10.1029/2008JD011470, 2009.

[22] K. Tanaka, T. Berntsen, J.S. Fuglestvedt, and K. Rypdal. Climate effects of emission standards: The case for gasoline and diesel cars. Environmental Science and Technology, 46(9):5205–5213, 2012.

[23] United Nations Environment Programme Finance Initiative. Why environmental externalities matter to institutional investors. URL: http://www.unpri.org/files/6728_ES_report_environmental_externalities.pdf, 2011. Accessed: August 20, 2012.

[24] Greenhouse Gas Protocol. GHG Protocol Product Life Cycle Accounting and Reporting Standard ICT Sector Guidance, URL: http://www.ghgprotocol.org/feature/ghg-protocol-product-life-cycle-accounting-and-reporting-standard-ict-sector-guidance. Accessed: August 20, 2012.

Biography

Anders S.G. Andrae received the M.Sc. degree in chemical engineering from the Royal Institute of Technology, Stockholm, Sweden, in 1997, and the Ph.D. degree in electronics production from Chalmers University of Technology, Gothenburg, Sweden, in 2005. He worked for Ericsson with LCA between 1997 and 2001. Between 2006 and 2008 he carried out post-doctoral studies at the National Institute of Advanced Industrial Science and Technology (AIST), Tsukuba, Japan. His specialty is the application of sustainability assessment methodologies to ICT solutions from cradle-to-grave. Dr. Andrae was recently the Editor of European Telecommunications Standards Institute (ETSI) first LCA standard for ICT. He has previously published three books, three theses, 19 conference papers, and 13 peer-reviewed journal papers. Since 2008 Dr. Andrae is with Huawei Technologies in Sweden as Senior Expert of Emission Reduction/Ecodesign/Sustainability/LCA.

Green Performance Indicators for Energy Aware IT Systems: Survey and Assessment

Sara S. Mahmoud and Imtiaz Ahmad

Computer Engineering Department, Kuwait University, Kuwait;
e-mail: imtiaz.ahmad@ku.edu.kw

Received 10 July 2012; Accepted: 26 September 2012

Abstract

Green computing provides techniques to reduce the wastage of energy, making it very critical to the development of IT systems due to the increasing power and energy needed now-a-days to run data centers. Green metrics play a vital role in green computing since they are criteria for evaluating the green performance of the large scale IT systems. Green computing metrics need to be defined to measure power costs and energy consumption. Measuring the amount of energy consumed by IT systems is a direct way of quantizing the amount of energy wasted or used efficiently by data centers. Energy awareness in applications and data centers can be obtained and calculated through green metrics such as the *Green Performance Indicators* (GPIs). The GPIs are classified into four classes: IT Resource Usage GPIs that compute resource usage, the Application Lifecycle KPIs that define efforts required to develop or redesign applications and reconfigure IT-infrastructure, the Energy Impact GPIs that represent the environmental impact of data centers, and the Organizational GPIs that describe organizational factors. Dividing the GPIs into four classes is an approach accepted in the EU Project GAMES. Many metrics have been identified by many associations, but there still yet to exist a framework in which a set of metrics are defined and used by applications and data centers to measure energy efficiency. This paper is a survey that sums up all the known metrics defined by almost all associations, and explicitly states their different units, scopes, and most importantly compares between

Journal of Green Engineering, Vol. 3, 33–69.

them to define similarities and replacements that may exist. Exact similarities may help eliminate metrics whose measurements can be accomplished by another metric. The gathered GPIs are defined into the four main classes mentioned above and frameworks built based on the similarities found in each class, are developed. We also define a framework for the relationship between green computing and GPIs to illustrate how GPIs are incorporated in the process. Since GPIs measure different factors related to energy consumption, they may have different units. We furthermore compare and critique different approaches to unify the various units of the different metrics, and implement one of the techniques on most of the metrics mentioned below with some improvements. We introduce the idea of correlations which can measure the amount of similarities found between the metrics to reduce the number of metrics to unify.

Keywords: Green Computing, Green Performance Indicators (GPI), Key Performance Service Indictors (KPI).

1 Introduction

The amount of energy needed to operate large scale systems such as the ever growing data centers is increasing. All data is sourced from or passes through a data center, making it the fundamental component in modern IT infrastructure. As the computing power of data centers grew so did the electricity usage causing an increase in heat dissipation, power consumption, and production. It is reported that power and cooling costs are the most dominant costs in data centers [4]. The disadvantage of the computing life cycle includes pollution in the form of carbon dioxide from power plants, and lead and mercury from manufacturing processes. For organizations with data centers to survive the boosting price of energy efficiency techniques must be implemented. Thus the study and idea of green computing is spreading in the branch of IT systems.

Green computing is a new approach which aims at designing computer systems that achieve better processing and performance with the least amount of power consumption [2]. There are many existing green computing techniques such as, reducing the overall power costs and developing energy aware and high performance computing systems traditionally, or virtualization technology, and waste recycling recently. To attain greener data centers, green computing metrics should be defined to measure their power usage and energy consumption. The metrics are criteria for evaluating the performance of data centers. Green performance metrics for data centers are a set of measurements that can qualitatively or quantitatively evaluate the environmental effects by operating a data center [4].

Green computing techniques can be implemented if needed, based on the values of the green performance metrics. Green Performance metrics play a key role in building new green applications and systems. Despite the importance of green metrics there has been quite little activity in collecting, summarizing, and agreeing on one set of performance metrics. There is no widely accepted metric set allowing for easy measuring and monitoring of energy consumed and wasted by applications and data centers, neither is there a clear method for applying these metrics. For the sake of simplicity, we will just use data centers instead of applications and data centers whenever we want to refer to both. Many associations defined different metrics, but they are not the same and if used improperly they may lead to contradictory conclusions. Thus a comprehensive and applicable framework to accurately measure energy efficiency and greenness of data centers is still scarce and not out there.

Recent works have been done such as in [2–6] presenting different frameworks for metrics associated with the different scopes of data centers but not all scopes. Tiwari [2], and Wang and Khan [4] for example, mainly mentioned metrics only associated with the energy/power. Kipp et al. [1], on the other hand, define green metrics into four classes that characterize the whole system view. The four class method defined in [1] is accepted and framed with the EU Project GAMES about green IT. But in [1], they only collect a small number of metrics that exist in each class, rather than mentioning almost all the metrics that are out there defined by many institutes. Other works have been presented to explain several green computing techniques like by Talebi and Way [10]. Fugini and Maestre [7] propose the idea of a *Green Certificate* which requires a unified unit for the several metrics to help compare the energy efficiency between different data centers and applications. Green computing has also been gaining a lot of attractions these days in works found in [8, 9, 12–14, 16–22]. This paper seeks to collect almost all the existing green metrics associated with all the scopes of a data center introduced and defined by many institutes such as the Green IT Promotion Council [30], the Uptime Institute [27], the Green Grid [26], the Nomura Research Institute [28], and the Emerson Corporation [29] and organizes each metric into one of four classes: IT resource usage, Application lifecycle, Energy impact, and Organizational. Each of these four classes comprises metrics that measure different parameters of different scopes in a computer system and data center. The four class method is based on the method proposed in [1]. The definition, unit, and scope of each metric are given along with comparisons with other metrics of the same class. This paper presents research, definition, and comparisons to find similarities, differences, and replacements among the various green performance metrics gathered. We then develop frameworks based on these comparisons. Another goal of this paper is to compare between different approaches to unify green metrics into one unit and to implement one of the techniques with some improvement. The difficulty of having many different metrics with many different units makes it a difficult task to define the "greenness" of an data center, and makes it difficult to compare between different applications built to perform the same functionality. Unifying the units for green performance indicators provides a standard for analyzing energy consumption in a comparative manner. Having a unified unit leads to implementing other ideas like green certificates as explained above. No such study exists that presents a survey of almost all the metrics associated with all the scopes of a data center defined by most associations, divides them into the four classes accepted in the EU Project GAMES, develops frameworks for the relationship between green computing and GPIs, and then compares between different approaches to unify the units of the different metrics along with critique and some improvements.

The rest of the paper is organized as follows. Section 2 explains the relationship between green computing and green performance metrics along with a framework. Section 3 defines GPIs in the IT Resource Usage class (first class). Section 4 defines key performance service indicators (KPIs) in the Application Lifecycle class (second class). Section 5 defines GPIs in the Energy Impact class (third class). Section 6 defines GPIs in the Organizational class (fourth class). Section 7 compares between different approaches to unify the units of different metrics with critique and includes our implementation on one of the techniques with some improvement. Section 8 concludes the paper and describes future work.

2 Green Computing and Green Performance Indicators

Green Computing is a discipline that studies, develops and promotes techniques for improving energy efficiency and reducing waste in the full life cycle of computing equipment from initial manufacture, through delivery, use, maintenance, recycling and disposal in an economically realistic way [10]. Through green computing, wastage of energy and power can be reduced by many techniques such as turning off the equipment when not in use. To achieve more energy efficiency and greener data centers, tools need to be defined to measure power costs and energy consumption. These tools are known as metrics [3]. Measuring the energy efficiency, power consumption, quality of the components deployed using various metrics can evaluate the environmental effects of a data center and its "greenness". The green performance indicators (GPIs) are a set of metrics defined to serve the purpose of measurement. The metrics cannot just identify and specify how green a data center is, they can also evaluate the products to compare similar data centers, track the "green" performance to increase efficiency, and provide guidance to engineers and service providers to develop research on future green data center technologies. Green performance metrics can quantitatively and qualitatively evaluate the environmental effects of IT systems. GPIs measure the level of greenness of applications and large IT systems.

Having a measurement is the best way to identify and comprehend the energy usage of a data center to point out easily where energy is being consumed the most and where it is wasted. Measurements also help in comparing and assessing the energy usage of a data center with others. Green Performance metrics are aimed at providing information that allows designers to provide better design decisions that work toward green systems. The GPIs framed within the EU Project Games about green IT; consider service applications and systems from the hardware usage, service life cycle, environmental perspectives, and organizational perspectives. These four considerations form four classes in which GPIs can be categorized into: IT Resource Usage GPIs, Application Lifecycle KPIs, Energy Impact GPIs, and Organizational GPIs. It is assumed that the functional part of an application has been decided and that the green metrics depend on the application structure (namely data flow, control flow, type of employed volatile object or database requirements) and on the specific hardware and software infrastructure configurations. GPIs contribute to green computing by measuring the energy efficiency of data centers. Then various green computing techniques can be implemented or adopted to "greenify" a data center if needed, based on the measurements taken by the GPIs.

We represent the relationship between the different phases of the green computing process and how GPIs are incorporated in these phases to contribute to green computing by giving measurable values of energy efficiency of a data center in Figure 1. This diagram represents the green computing process in five different phases resulting in a green data center. The five phases begin with specifications, then design, implementation and usage, recycling and disposal, and finally analysis. Each phase is described next in detail. GPIs

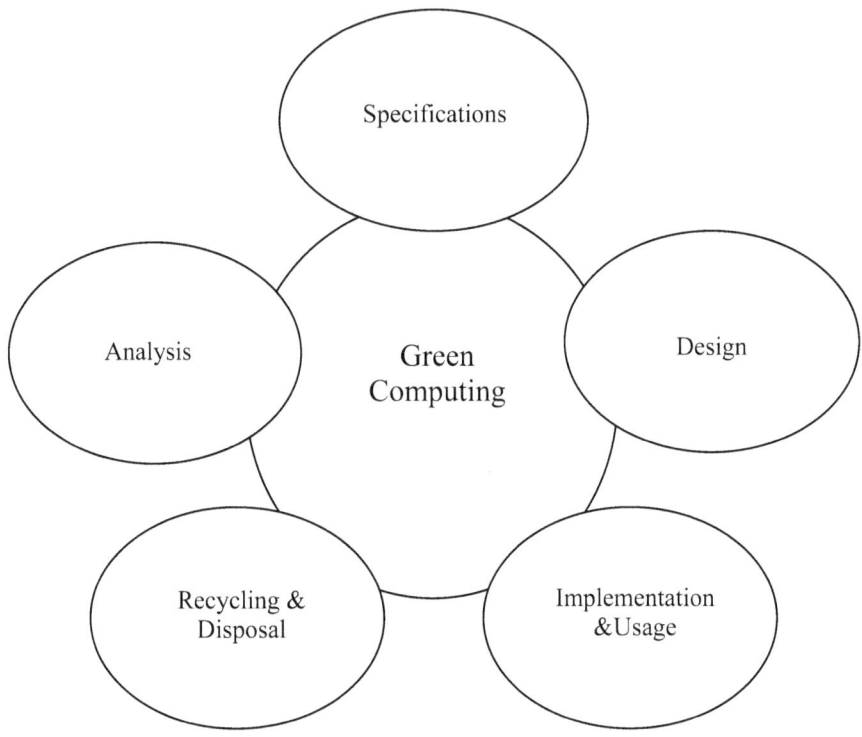

Figure 1 Green computing process.

are incorporated in all the phases of green computing to measure efficiency and performance, and to determine if greener solutions need to be adopted.

2.1 Specifications

The first phase involved in the green computing process is the specification stage. This phase identifies current green computing initiatives, best practices for implementing green data centers [24], and constraints. Green computing initiatives include buying greener office supplies and furniture, encouraging green-friendly transportation, increasing waste recycling programs, conserving water, buying wind or solar powered electricity, buying carbon offsets, installing air and water filtration equipment, and using green products in building renovations and new construction. Constraints are guidelines for managing data centers according to eco-related laws and regulations. Thus

GPIs play a role in the specification phase because they indicate the level of energy efficiency that is preferable.

2.2 Design

The second phase of the green computing process is design. This phase defines how to design an energy effect data centre. "Green" elements should be involved in the design process of a data center. Energy efficient data center design should address all of the energy use aspects included in a data center: from the IT equipment to the HVAC equipment to the actual location, configuration and construction of the building [23]. There are five primary areas to which energy efficient design practices can be implemented. These five areas are: IT system, Environment Conditions, Air Management, Cooling Systems, and Electrical System [23]. Green design decisions are effective at reducing environmental burdens and reducing costs as compared to the process of ignoring environmental effects during the design stage and ending up using clean up strategies. Some green design decisions include solvent substitution in which a toxic solvent is replaced with a benign alternative, or technology change such as more energy efficient semi-conductors. The Energy star program specifies maximum energy consumption standards for computers and other electronic devices [11]. GPIs definitely have an impact on decision making through classifying the efficiency of the elements used and the choices taken.

2.3 Implementation and Usage

The third phase of the green computing process is implementation and usage. This phase describes the implementation of green design decisions and usage of IT systems such computers, servers, and associated subsystems. These include such as monitors, printers, storage devices, and networking and communications systems. For instance, computer usage may generate a great deal of paper. User may minimize unnecessary waste by double-checking documents for accuracy before printing. Another useful way to practice green computing usage is by engaging power management features on the computer to allow screens and hard drives to become inactive after minutes of being idle. Operating Energy Star labeled equipment contributes to green computing implementation. A major part of this phase is applying metrics to measure efficiency of all the aspects in the data center.

2.4 Recycling and Disposal

The fourth phase of the green computing process is recycling and disposal. This phase covers the disposal or recycling of data center equipment at the end of its lifecycle in an environmentally responsible fashion. Like all other equipment, data center equipments are manufactured, sold, used, and often reused, and then ultimately disposed of [24]. A disposal may mean the equipment is discarded/destroyed or sold to be used again by another organization. Data centers managers replace the data center equipment either by regular refresh cycles, or wait till they have to, or utilize a continuous update process. The recycling policy should follow the 3R's policy (reduce, reuse, recycle) for proper recycling of the data center equipment. Reducing is associated with buying only what is needed which results in fewer good being produced and less throwing away of products. Reuse is associated with buying products that can be repeatedly used resulting in saving natural resources since the product is not thrown away. Recycling is associated with recycling materials leasing to reduction in energy needs for mining, refining, and other manufacturing processes.

2.5 Analysis

The fifth phase of the green computing process is analysis. This phase is related to measuring the performance of a data center regularly using metrics selected [24]. The steps associated to this phase begin with collecting data at regular intervals, then performing analysis energy metrics, comparing new values with old values, and finally look for greener solutions and continue the greening process.

Figure 2 shows our arrangement of the various green performance metrics (GPIs) categorized into four classes accepted by the EU Project Games. Each class contains two or more types of metrics. For example under the IT resource usage GPIs are metrics that either measure the utilization and efficiency of the equipment or metrics that measure the space efficiency of a data center based on the IT equipment deployed. In Sections 3 to 6, all the metrics related to the classes are discussed.

3 IT Resource Usage Metrics

The first class of Green Performance Indicators measures the energy consumption of an application/data center. The energy consumption varies on the type of application running. Some applications run the processor intens-

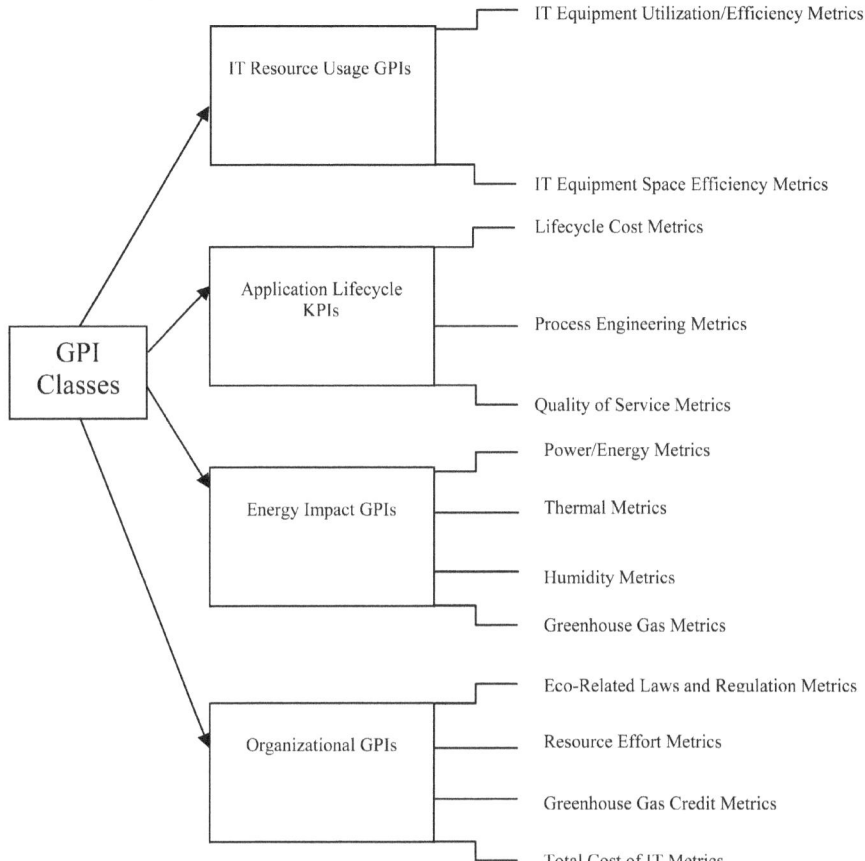

Figure 2 Green performance networks arrangement.

ively, others are data intensive, while others may even be a hybrid of both. Thus the energy consumption of an application is defined as a function of its resource utilization. These resources include CPUs, servers, storage devices, main memory, and I/O resources.

A suggested framework for the metrics found in the IT resource GPIs class is shown in Figure 3. Each metric is either an IT equipment/utilization metric or an IT equipment space efficiency metric. Metrics that have similarities between each other or may replace each other are grouped together in one rectangle. These similarities and replacements between the metrics, if any for IT resource GPIs are all found in Tables 1 and 2. Note the metrics in the IT equipment utilization/efficiency section are dimensionless (percentages). On

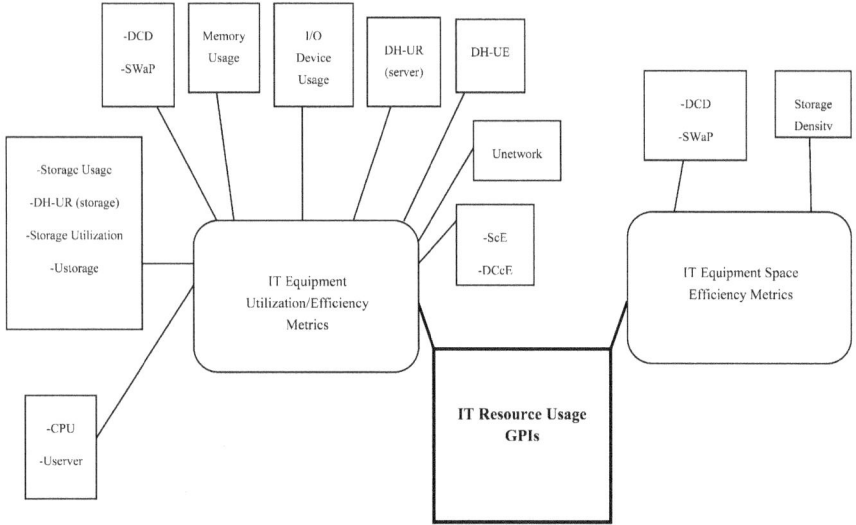

Figure 3 Framework for IT resource GPIs.

the other hand the dimensions of the IT equipment space efficiency section have dimensions. Most of the metrics in this category are defined in the rest of this section.

CPU usage: This metric refers to the amount of CPU utilization needed to process the instructions of an application. It is measured by evaluating the percentage of time that the allocated CPU spends on performing computing operations [1].

Memory usage: This metric refers to the amount of main memory (RAM) usage. It is measured by evaluating the percentage of RAM allocated for a specific application [1].

I/O device usage: I/O operations allow applications to communicate with system devices. This metrics refers to the occupation of an I/O device. It is measured by evaluating the percentage of occupation of a resultant I/O device for communication and the number of messages transferred by an application over a set of system components [1].

Storage usage: This metric refers to the amount of storage utilization for data read and write to a permanent storage such as local or remote hard disk drives [1]. It is measured by evaluating the entire storage utilization for an application.

These four resource usage GPIs can be used to measure energy consumption on different levels such as a single computer, on a class level, or for an entire IT service centre [1].

Deployed Hardware Utilization Ratio (DH-UR): This metric is an indicator for the energy consumption of the main computing equipments from the IT Service Centre. This metric specified e-waste reduction approaches by removing or suspending IT equipment such as servers or storage within the data centre [3]. This metric is defined by the Uptime Institute.

- DH-UR (server) = Number of servers running live application/Total number of servers actually deployed.
- DH-UR (storage) = Number of terabyte of storage holding important frequently accessed data (within at least 90 days)/Total terabyte of storage actually deployed.

Deployed Hardware Utilization Efficiency (DH-UE): This metric helps measure the potential improvement in energy savings by the utilization of servers and storage via virtualization [3]. Knowing the load of servers and the total number of servers helps in finding out how much energy is consumed during peak loads relative the number of servers being used actively. This metric is defined by the Uptime Institute.

DH-UE = Minimum number of servers necessary to handle peak compute load/Total number of servers deployed

Storage Density (SD): This metric is defined by the Nomura Research Institute.

SD = Storage Utilization/Total Data Center Square Footage

Storage Utilization (SU): This metric is defined by the Nomura Research Institute.

SU = Server Network and Backup Storage in Use/Total Storage Available

Userver: The Server utilization metric measures the rate of maximum ability of the processor in the highest frequency state [3]. This metric is defined by the Green Grid.

Userver = Activity of the server's processor/Maximum ability in the highest frequency state

Ustorage: The Storage utilization metric defines a ratio through which percentage of used storage regarding to overall storage capacity will be measured within the data center [3]. This metric is defined by the Green Grid.

Ustorage = Percent storage used/Total storage capacity of data center

Unetwork: The Network utilization metric depicts the percentage of used bandwidth over total bandwidth capacity within the data center [3]. This metric is defined by the Green Grid.

Unetwork = Percent network bandwidth used/Total bandwidth capacity of data center

Server Compute Efficiency (ScE): This metric measures the efficiency of servers in data centers over anytime period by summing the number of samples where the server is found to be providing primary services (p) and dividing this by the total number of sample (n) taken over that time period and multiplying by 100 [2]. This metric helps managers improve total energy by determining the servers that are not providing primary services for specific periods. These servers can be switched off or even virtualized. This metric is defined by the Green Grid.

$$ScE = \left(\sum_{i=1}^{n} P_i/n \right) \times 100$$

Data Center Density (DCD): This metric quantifies the data center space efficiency [1]. It is defined by the Green Grid Institute.

DCD = (Power of all Equipment)/(Data Center Space Area)

Space, Watts, and Performance (SWaP): This metric gives users an effective cross comparison and total view of a server's overall efficiency [4]. It characterizes a data center's energy efficiency using three parameters.

SWaP = Performance/(Space × Power Consumption) where Performance is based on industry standard benchmarks, Space is the measurement of the height of the server in rack units, and Power is measured by determining the watts consumed by the system.

Data Center Compute Efficiency (DCcE): This metric does not measure how much work is done, instead how much work is useful. It offers a track system that enables a data center operator to calculate efficiency of computing in servers and decide on the appropriate number of servers needed to do the job at hand [3]. This metric is defined by the Green Grid.

$$DCcE = \sum_{j=1}^{m} ScE/m$$

Table 1 IT equipment utilization/efficiency GPIs.

Metric	Definition	Unit	Scope	Similarities/Replacements
CPU usage	The amount of CPU utilization needed to process instructions	Percentage	Application/Data Center	The Userver is similar to this metric because it also measures CPU utilization. Either one of them can replace the other.
Memory Usage	The amount of main memory (RAM) usage	Percentage	Application/Data Center	
I/O Device Usage	The occupation of an I/O device	Percentage	Application/Data Center	
Storage Usage	The amount of storage utilization for data read and write	Percentage	Application/Data Center	This metric can be replaced by the DH-UR(storage) metric.
DH-UR (server)	Indicator for the energy consumption of servers	Percentage	Application	
DH-UR (storage)	Indicator for the energy consumption of storage	Percentage	Application	This metric can replace the Storage Usage, the Storage Utilization, and the Ustorage metrics. This metric not only measures the entire storage utilization, instead it also measures the number of terabyte storage in 90 days to suspend the storage that is not being used.
DH-UE	Measures the improvement in energy savings due to utilization of servers and storage via virtualization	Percentage	Data Center	
Storage Utilization	Utilization of Storage	Percentage	Application/Data Center	This metric can be replaced by the DH-UR(storage) metric.
Userver	Rate of maximum ability of the processor	Percentage	Application/Data Center	This metric is similar to the CPU usage metric because it also measures CPU utilization. Either one of them can replace the other.
Ustorage	Percentage of used storage regarding to overall storage capacity	Percentage	Application/Data Center	This metric can be replaced by the DH-UR(storage) metric.
Unetwork	Percentage of used bandwidth over total bandwidth capacity	Percentage	Application/Data Center	
ScE	Measures the efficiency of servers in data centers over anytime period	Percentage	Application/Data Center	Both this metric and the DCcE metric define the usefulness of the work being performed.
DCcE	Measures how much work is useful.	Percentage	Data Center	Both this metric and the ScE metric define the usefulness of the work being performed. This metric aggregates the ScE metric across all servers in a data center as well.

In Tables 1 and 2 a list of all the metrics in class one are listed with their specified unit, scope in which they may be used, any similarities and differences that might exist between one metric and another. Finding similarities and differences helps compare between the different metrics. Exact similarities may help eliminate metrics whose measurements can be accomplished by another metric.

Table 2 Equipment space efficiency metrics.

Metric	Definition	Unit	Scope	Similarities/Replacements
DCD	Quantifies the data center space efficiency	KWatt/squa re footage	Data Center	This metric is similar to SWaP in that is quantifies the data center's space efficiency.
SWaP	This metric gives users an effective cross comparison and total view of a server's overall efficiency.	Operations/ (RU* Watts)	Data Center	This metric is similar to DCD in that is quantifies the data center's space efficiency, but this metric also allows for the comparison of IT server configurations.
Storage Density	Density of Storage	Percentage/ (m/cm^3)	Data Center	

4 Application Lifecycle KPIs

The metrics in this class are quite different from the ones found in classes one, three, and four. These metrics do not directly impact the energy consumption of data centers, instead they are used to determine the performance of applications. The metrics in this class characterize process quality and efforts for designing and maintaining the process. For this reason the metrics in this class are named Key Performance Service Indicators instead of Green Performance Indicators. The metrics in this class are categorized into three categories. The first category is the Lifecycle Cost indicators, the second is the Process Engineering category, and the last is the Quality of Service category. The metrics of the three categories are explained next.

In Figure 4, is our suggested framework for the metrics found in the Application Lifecycle class. Each metric is either a Lifecycle cost metric or a QoS metric or a process engineering metric. Metrics that have similarities between each other or may replace each other are found together in one rectangle. There are no relationships found between the metrics in this class.

4.1 Lifecycle Cost Indicators

Lifecycle cost indicators or metrics describe the total process lifecycle expenses. The scope of these metrics is the application being built. These metrics include:

1. Cost of conceptual modelling
2. Cost of analysis
3. Cost of design
4. Cost of development
5. Cost of deployment

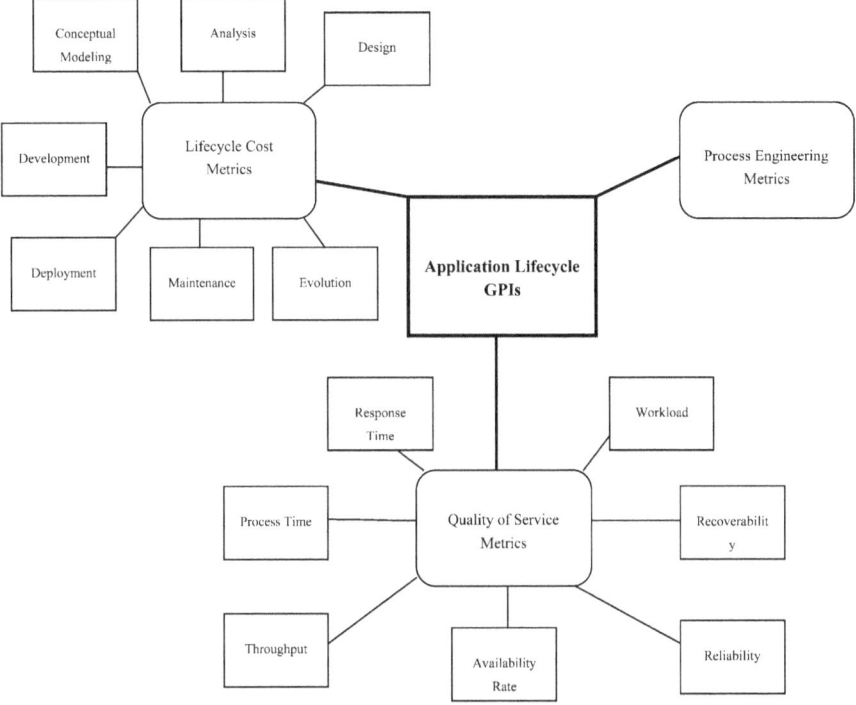

Figure 4 Framework for application lifecycle GPIs.

6. Cost of maintenance
7. Cost of evolution

The above metrics take into account potential parameters such as the developer's experience, the main service operation complexity, the level of abstraction, the reusability and integration rates, the required stability, and the closeness of the application to the business core. If we consider the cost of development metric, the parameters that are considered are the efforts placed by the developers during the application lifecycle. More precisely, if the application runs are observed throughout time it can be determined how an application should be redesigned to be more energy efficient. "The cost for redesign with respect to shorter executions, storage savings, and other application parameters create an index of energy saving" [1]. Thus the indicators in this category are given in energy measures.

4.2 Process Engineering

The indictors in this category illustrate the style of development used during design, coding, and deployment of an application. The scope of these metrics is the application being built. Parameters that affect the style of deployment include the level of maturity of the used development platform regarding the tools for coding, documentation and the engineering methods. Such parameters have an impact of energy consumption by enabling the developers to build applications faster with fewer errors, efficient coding, and built in elements that reveal energy leakages. Parameters that affect the style of code include indexes of data usage, service usage and branching probabilities. For example, the less number of times an application runs in dead branch flows, the more energy efficient it will be.

4.3 Quality of Service

The metrics in this class evaluate the quality measures of an application. The scope of these metrics is the application being built. These metrics have an indirect instead of a direct impact on the energy consumption of an application. An increased reliability or response time will result in a negative impact on energy consumption. This expected quality can be instead achieved by efficient resource utilization. These metrics include:

Response Time: This metric refers to the time it takes a service to handle a user's request. The response time given Tresponse of a given service S is calculated using the processing time Tprocess and the transmission time Ttrans [1]. The unit for this metric is time.

$$\text{Tresponse} = \text{Tprocress}(S) + \text{Trans}(S)$$

Process Time: This metric is given by the average time taken to process a service S from the time of invocation Tinvocation to the time of completion Tcompletion including delays Tdelay [1]. The unit for this metric is time.

$$\text{Tprocess}(S) = (\text{Tcompletion}(S) - \text{Tinvocation}(S)) + \sum \text{Tdelay}$$

Throughput: This metric is the average number of service requests successfully served during a given period of time [1].

Availability Rate: The average rate of availability of a given service S is represented by the probability that a certain request is properly responded within a maximum expected time frame [1]. The unit for this metric is percentage.

Availability Rate=number of successful service requests/number of service requests

Reliability: This metric is given by the probability that a system remains operational over a certain period of time [1]. It can be represented by the exponential distribution that describes random failures.

$R = e^{[-(1*t)]}$ where t is the expected execution time and l is the failure rate over the reference interval

Recoverability: Recoverability is given as the probability that a failed system is can be operational again within a certain period of time [1]. It can be measured from logs using process mining. The unit for this metric is percentage.

Workload: This metric includes the type and rate of requests, execution of the software packages, and in-house application programs sent to the system [1]. The unit for this metric is given by an index.

5 Energy Impact GPIs

The GPIs of the third class describe the impact if IT service centers and applications on the environment. Metrics in this class measure the amount of power supply needed by a data center, consumed materials, CO_2 emissions, and other energy related factors rereleased in the air. These metrics give a direct value of how "green" an application or data center is.

In Figure 5, is our suggested framework for the metrics found in the Energy Impact class. Each metric is either a power/energy metric or a thermal metric or humidity metric or a greenhouse gas metric. Metrics that have similarities between each other or may replace each other are found together in one rectangle. These similarities and replacements between the metrics, if any for Energy Impact GPIs are all found in Tables 3, 4, 5 and 6. In the power/energy section, a connection if found between the PUE metric and all the metrics in this rectangle. The same case is found between the ITEE metric and the rest of the metrics in that rectangle for the same section.

IT Equipment Utilization (ITEU): Is a metric that defines the energy-saving level through implementing both virtual and operational techniques among IT equipments in a data centre [3]. It measures the average utilization of the entire IT equipment within a data center. This metric is defined by the Green IT Promotion Council.

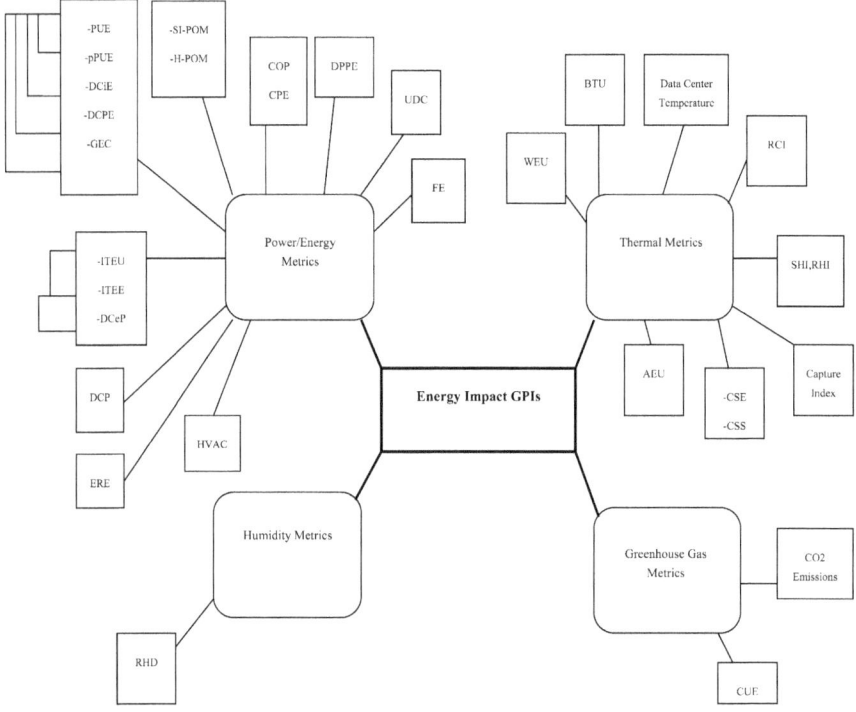

Figure 5 Framework for energy impact GPIs.

ITEU = Total measured power of IT equipment/Total rated power of IT equipment

Data Centre Infrastructure Efficiency (DCiE): It is used to determine the energy efficiency of a data centre. This metric refers to how much energy the IT equipment consumes from the total energy consumption [1]. This metric is defined by the Nomura Research Institute.

DCiE = IT Equipment Power/Total Facility Power, where IT equipment power is defined as the load associated with computers, storage, and network equipment and Total Facility power is measured at or near the facility utility meter

Power Usage Effectiveness (PUE): This is a metric that focuses on the data center infrastructure. Its value may range from 1.0 to infinity. A value 1.0 indicates 100% efficiency [2]. This metric is defined by the Green IT Promotion Council.

PUE = Total Facility Power/IT Equipment Power

Partial Power Usage Effectiveness (pPUE): This is a conceptual metric where a PUE-like value for a subsystem is measured and reported [2]. This metric is defined by the Green Grid.

pPUE = Total Energy within a boundary/IT Equipment Energy within that boundary

Site Infrastructure Power Overhead Multiplier (SI-POM): This metric defines how much power is consumed in overhead instead of critical IT equipments [3]. This metric is defined by the Uptime Institute.

SI-POM = Data center power consumption at the utility meter/Total hardware AC power consumption at the plug for all IT equipment

IT Hardware Power Overhead Multiplier (H-POM): This metric defines how much power input to hardware is wasted in power supply for fans rather than useful computing components [3]. H-POM differs for a single device and for an entire data centre. This metric is defined by the Uptime Institute.

H-POM (single device) = AC hardware load at the plug/DC hardware Compute load
H-POM (data centre) = Total hardware load at the plug for the entire data centre/total hardware compute load for the entire data centre.

Data Centre Energy Productivity (DCeP): This metric indicates the number of bytes which are processed per kWh of electric energy. This metric is capable of measuring site infrastructure and IT equipment while assessing data center efficiency [3]. This metric is defined by the Green Grid.

DCeP = Useful work produced in a data center/Total energy consumed in the data center to produce that work

Data Centre Performance Efficiency (DCPE): This metric shows how effective a data centre is in terms of power consumption when work is given [3]. Because this metric is emerging over time, it is complicated to measure. This metric is defined by the Green Grid.

DCPE = Useful Work/Total Facility Power

Coefficient of Performance of the Ensemble (COP): This metric measures the IT data centre greenness. "It reflects the energy efficiency of the data centre cooling system by taking into account the data centre performance per unit of used energy" [1].

COP Ensemble = Total Heat Dissipation/(Flow Work + Thermodynamic Work) of cooling system

CO_2 *Emission*: This metric is the amount of average carbon dioxide emissions from generating an average kWh of electricity [1].

Compute Power Efficiency (CPE): This metric seeks to quantify the overall efficiency of a data center. It considers that not all electrical power delivered to the IT equipment is transformed by that equipment into useful work.

CPE = (IT Equipment Utilization $\times$ IT Equipment Power)/Total Facility Power
or CPE = IT Equipment Utilization/PUE

Energy Reuse Effectiveness (ERE): It is a metric that measures the energy efficiency in data centers that re-use waste energy from their own data center [2]. This is a metric that focuses on recycling and reusing of components. This metric is defined by the Emerson Corporation.

ERE = (Cooling + Power + Lighting + IT-Reuse)/IT where IT is the energy used by all of the IT equipment (servers, network, storage) in the data center

Carbon Usage Effectiveness (CUE): This metric measures sustainability of data centers [3]. It can help organizations recognize whether their current data centers are efficient before they decide to implement a new one. This metric is defined by the Greed Grid.

CUE = Total Carbon Dioxide Emissions from Total Data Center Energy/IT Equipment Energy

Water Usage Effectiveness (WUE): This metric is defined by the Green Grid.

WUE = Annual Site Water Usage/IT Equipment Energy

IT Equipment Energy Efficiency (ITEE): This metric improves energy saving through setting up new equipments with high processing capacity in term of power consumption [3]. This metric is defined by the Green IT Promotion Council.

ITEE = Total server capacity + total storage capacity + total NW equipment capacity/Rated power of IT equipment

Green Energy Coefficient (GEC): This metric is meant to persuade operators to use renewable energy [3]. This metric replaces grid electricity with green energy. It is a ratio that originated from dividing the value of Green Energy used in a data center by the total power consumed in the data center. This metric is defined by the Green IT Promotion Council.

GEC = Green Energy/DC total power consumption

Data Center Performance Per Energy (DPPE): This is an integrated metric created to improve energy savings in data centers [3]. It contains four metrics already explained above. This metric is defined by the Green IT Promotion Council.

DPPE = ITEU × ITEE × 1/PUE × 1/1-GEC

Facility Efficiency (FE): This metric is defined by the Nomura Research Institute.

FE = Facility Energy Efficiency × Facility Utilization

Data Center Utilization (UDC): This metric calculates the amount of power that IT equipment consume regarding to the data center's real capacity [3]. This metric is defined by the Green Grid.

UDC = IT Equipment Power/Actual Power Capacity of the Data Center

Data Center Productivity (DCP): This metric is defined by the Nomura Research Institute.

DCP = Useful Computing Work/Total Facility Power

Relative Humidity Difference (RHD): This metric aims at measuring the amount of humidity in the air. Humidity is the measurement of moisture content in the air. High humidity in data centers result in hardware failures and increase the cooling costs making humidity control an essential for physical media. Humidity is measured by looking at the relative humidity which is given as a percentage and measures the amount of water in the air at a given temperature compared to the maximum amount of water that air can hold [4].

RHD = Rhumidity – Shumidity where Rhumidity is the return air relative humidity and Shumidity is the supply air relative humidity

Heating, Ventilation, and Air Conditioning (HVAC) Effectiveness: The HVAC system of a data center includes the computer room air conditioning and ventilation, a central cooling plant, and minor load (lighting). The HVAC system effectiveness is the fraction of the IT equipment energy to the HVAC system energy [4].

HVAC Effectiveness = IT/(HVAC + (Fuel + Steam + Chilled Water) × 293) where IT is the annual IT Electrical Energy Use, and HVAC, Fuel, Stream, Chilled Water are all in terms of years

Another set of metrics that may be included in this section or class is thermal metrics. "It is reported that cooling costs can be up to 50% of the total energy cost in a data center" [3]. Thus thermal metrics are essential for operating a green data center.

Data Center Temperature: High temperature in a data center has a negative impact on the computing systems causing for reduction in reliability, quality of service, and longevity of components. It is recommended that IT equipment not be operating in a room temperature above 85°F (30°C). A small temperature differential between supply and return air temperature allows for an improvement in air management, the rise of supply air temperature, and thus reduction in energy usage [4].

British Thermal Unit (BTU): "A BTU is the amount of energy required to raise the temperature of a pound of water 1°F" [4]. To calculate the amount of cooling power needed to cool a data center there a three issues to consider:

- Size of a data center (determines how many BTUs are required to cool down a data center) – BTU = 330 × Length × Width
- Equipment (BTUs required for equipment in a data center) – BTU = 3.5 × total wattage running the equipments
- Lighting (total BTUs required for lighting) – BTU = 4.25 × wattage of lighting

Rack Cooling Index (RCI): This metric measures the effective coolness and maintaining of equipment racks with industry thermal guidelines and standards [4].

RCI_{HI} = This metric measures the absence of over-temperatures or characterize the equipment health at the high end of the temperature range

RCI_{LO} = This metric gives a measurement of the supply conditions when they are below the minimum recommended temperature. If the supply conditions of the equipment racks are below the recommended temperature then the humidity level may be harmful

Supply Heat Index and Return Index (SHI, RHI): These two measurements assess the magnitude of recirculation and blend of hot and cold streams. Best results occur when RHI is high and SHI is low [4].

- SHI = This metric measures the extent to which warm return air mixes with cool supply air
- RHI = This metric measures the extent to which cool supply air mixes with warm return air.

Capture Index (CI): This metric measure the cooling performance based on the airflow patterns related to the supply of cool air or the removal of hot air from the equipment rack. There are two CI metrics. The cold-aisle CI is the amount of air from local cooling resources ingested by the rack. The hot-aisle CI is the amount of air captures by local extract exhausted by the rack [4].

Data Center Cooling System Efficiency (CSE): This metric describes the overall efficiency of the cooling system in terms of energy input per unit of cooling output [4].

CSE = Average cooling system power usage/Average cooling load

Cooling System Sizing (CSS): This metric may indicate if a cooling system is scalable and has good potential. It is the ration of the installed cooling capacity to the peak cooling load [4].

CSS = Installed Chiller Capacity/Peak Chiller Load

Air Economizer Utilization (AEU): The metric indicates the extent to which an air-side economizer system is used to provide free cooling. It is described as the percentage of hours in a year that the economizer system is in full operation [4].

AEU = Air economizer hours/24 × 365

In Tables 3, 4, 5, and 6 all the metrics in the third class are listed with a brief definition, their units, scope, and any similarities, differences, or replacements.

6 Organizational GPIs

Organizational GPIs are the fourth and last class of GPIs. The metrics in this class measure organizational factors. Organizational GPIs have a direct impact on high level decisions in data centers related to infrastructural costs and guidelines defined by eco-related laws and regulations for managing data centers. Infrastructural costs are costs for owning new energy saving software and hardware, or maintenance, etc.

Table 3 Power/energy metrics.

Metric	Definition	Unit	Scope	Similarities/Replacements
ITEU	Defines energy-savings due to implementing virtual and operational techniques.	Percentage	Data Center	This metric and the ITEE metric reduce energy consumption of IT equipments in data centers.
DCiE	Determines the energy efficiency of a data centre.	Percentage	Data Center	This metric and the PUE both identify the data center's energy consumption and are predecessor requirements for green measurements.
PUE	This metric focuses on the data center's infrastructure.	Percentage	Data Center	This metric and the GEC metric reduce energy consumption of facilities in data centers. This metric and the DCiE both identify the data center's energy consumption and are predecessor requirements for green measurements
pPUE	PUE-like value for a subsystem.	Percentage	Data Center	This metric is exactly similar to the PUE metric except that is it limited to a subsystem.
SI-POM	Measures how much power is consumed in overhead in the site infrastructure.	Percentage	Data Center	This metric and the H-POM metric (single device/data center) measure the overhead power. This metric measures it in the site infrastructure as compared to the H-POM metric which measure the overhead in the IT equipment.
H-POM (single device)	Measure how much power input to hardware is wasted in power supply for non-computing components.	Percentage	Device	This metric and the H-POM metric for a data center, both measure the overhead power in IT equipment.
H-POM (data center)	Measure how much power input to hardware is wasted in power supply for non-computing components.	Percentage	Data Center	This metric and the H-POM metric for a single device, both measure the overhead power in IT equipment.
DCeP	Indicates the number of bytes which are processed per kWh of electric energy.	Tasks/kWhr	Data Center	This metric can be replaced by its equivalent metric ITEE.
DCPE	Measures how effective a data centre is using power to provide a given level of service.	Percentage	Data Center	This metric is an expansion of PUE. It is a refined version of PUE because is adopts all major power-consuming subsystems found in a data center.
COP	Measures the IT data centre greenness.	Dimensionless	Data Center	This metric and the CPE measure the overall efficiency or greenness on the data center. The COP metric measures the overall efficiency with respect to the cooling system and the CPE metric measures the overall efficiency with respect IT equipment.
CPE	Quantifies the overall efficiency of a data center.	Percentage	Data Center	This metric and the COP measure the overall efficiency or greenness on the data center. The COP metric measures the overall efficiency with respect to the cooling system and the CPE metric measures the overall efficiency with respect IT equipment.
ERE	This metric that focuses on recycling and reusing of components.	Percentage	Data Center	
ITEE	This metric improves energy saving through setting up new equipments with high processing.	DEC	Data Center	This metric can be replaced by its equivalent metric DCeP. This metric and the ITEE metric reduce energy consumption of IT equipments in data centers.
GEC	It persuades operators to use renewable energy.	Percentage	Data Center	This metric and the PUE metric reduce energy consumption of facilities in Data Centers.

Table 3 Continued.

DPPE	It is an integrated metric created to improve energy savings in data centers.	Operations/kWh	Data Center	
FE	Facility Efficiency	Joules	Data Center	
UDC	Calculates the amount of power that IT equipment consume regarding to the data center's real capacity	Percentage	Data Center	
DCP	Data Center Productivity	Joules/Watts=Time	Data Center	This metric and the DCcE metric both compute the amount of useful work. This metric measures useful work in all the data center
HVAC	Heating, Ventilation, and Air Conditioning	Percentage	Data Center	

Table 4 Thermal metrics.

Metric	Definition	Unit	Scope	Similarities/Replacements
RCI-RCI$_{HI}$, RCI$_{LO}$	Measures the effective coolness and maintaining of equipment racks	Percentage	Data Center	
SHI, RHI	Assess the magnitude of recirculation and blend of hot and cold streams.	Index	Data Center	
Capture Index-Cold-aisle CI, Hot-aisle CI	Measures the cooling performance based on the airflow patterns related to the supply of cool air or the removal of hot air from the equipment rack	Index	Data Center	
CSE	Describes the overall efficiency of the cooling system	kW/ton	Data Center	The metric along with the CSS metric measure the efficiency of the cooling system installed.
CSS	Indicate if a cooling system is scalable and has good potential	Percentage	Data Center	The metric along with the CSE metric measure the efficiency of the cooling system installed.
AEU	Iindicates the extent to which an air-side economizer system is used to provide free cooling	Percentage hours	Data Center	
WUE	Water Usage Effectiveness	Liters/kWh	Data Center	
Data Center Temperature	Measures the data center's temperature	Degrees	Data Center	
BTU	This metrics presents the amount of energy required to raise the temperature of a pound of water to 1˚F	Dimensionless	Data Center	

Table 5 Greenhouse gas metrics.

Metric	Definition	Unit	Scope	Similarities/Replacements
CO2 Emission	Measures the amount of average carbon dioxide emissions.	Percentage	Data Center	
CUE	This metric measures sustainability of data centers.	$Kg(CO_2)/kWh$	Data Center	

Table 6 Humidity metrics.

Metric	Definition	Unit	Scope	Similarities/Replacements
RHD	Measures the amount of humidity in the air.	Percentage	Data Center	

In Figure 6, is our suggested framework for the metrics found in the Organizational class. Each metric is either eco-related laws and regulations metric or a resource effort metric or a greenhouse gas credit metric or a green solution metric or a cost of the IT center metric. Metrics that have similarities between each other or may replace each other are found together in one rectangle. These similarities and replacements between the metrics, if any for Organizational GPIs are all found in Table 7.

Human Resources Indicator: This metric assesses efforts spent by human resources involved in running and managing an application, service development [1].

Compliance Indicator: This metric evaluates the amount of efforts spent abiding with the government regulations and/or consortium policies [1]. Some consortium policies include EU Code of Conduct for Data Centers 2010.

Infrastructural Costs Indicator: This metric includes the costs related to the building, the facility, and IT equipments [1]. This metric also involves the costs related to replacement and maintenance of the IT equipments. Costs/efforts of humans who design, run, and manage the data center as a whole are also included.

Carbon Credit: This metric represents the offset credits that are bought and sold to offset carbon dioxide emissions [1]. This rules and conditions for this metric differ from country to country.

Return of Green Investment (RoGI): This metric measures the time it takes for green solutions to pay off or recuperate [1].

Consumables Index: This metrics is associated with costs of printouts and materials produced by the service during executions [7].

Total Cost of Ownership: This is a metric that represents the cost it takes an owner to purchase or build, operate, and maintain a data center. The

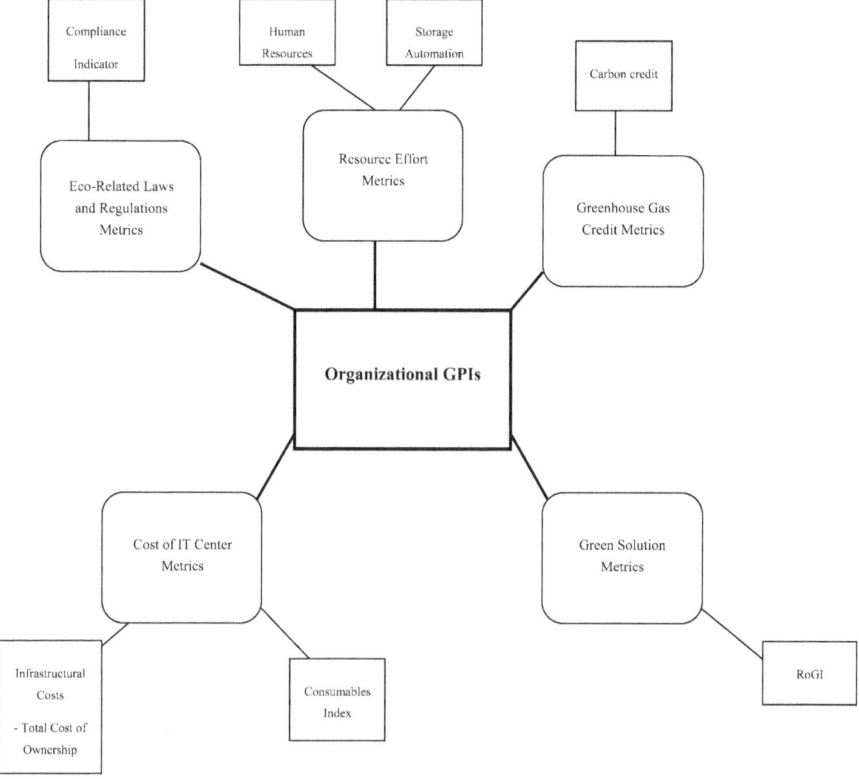

Figure 6 Framework for organizational GPIs.

total expenses are divided into two parts which the capital expenses and the operational expenses [4]:

- Capital Expenses: These are expenses related to the initial investments for purchasing and building a data center. The major subsystems of a data center such as the cooling and space are scale almost linearly with the amount of energy consumed. Thus the capital cost is represented in dollars per watt.
- Operational Expenses: These are the monthly expenses of running a data center. Factors that affect these expenses are climate changes, management costs, or even implementation techniques.

Storage Automation: Human Operators/Storage Density where Storage Density = Storage Utilization/Total Data Center Square Footage [3].

Table 7 Organizational GPIs.

Eco-related Laws and Regulations			
Definition	Unit	Scope	Similarities/Replacements
Amount of efforts spent abiding with the government regulations and/or consortium policies	Index	Data Center	
Resource Effort Metrics			
Efforts spent by human resources	Costs	Application/Data Center	
Storage Automation	Percentage	Data Center	
Greenhouse Gas Credit Metrics			
Offset credits that are bought and sold to offset carbon dioxide emissions	Offset Credits	Data Center	
Green Solution Metrics			
The time it takes for green solutions to pay off or recuperate	Time	Data Center	
Cost of IT Center Metrics			
Costs related to the building, the facility, and IT equipments and costs/efforts of humans who design, run, and manage the data center	Cost	Data Center	This metric is similar to the Total Cost of Ownership metric because both refer to the costs of building, operating, and managing the data center. But this metric cannot replace the Total Cost of Ownership metric because, it also considers human efforts that run the data center.
Represents the cost it takes an owner to purchase or build, operate, and maintain a data center	Capital Costs and Operational Costs	Data Center	This metric is similar to the Infrastructural cost metric because both refer to the costs of building, operating, and managing the data center.

In Table 7 all the metrics in the fourth class are listed with a brief definition, their units, scope, and any comparisons, or replacements.

7 Approaches to the Unification of Different Units of GPIs

In this section we define three different recent techniques found for unifying the different units of the green metrics to simplify the task of comparing data centers and application by having one value to represent the energy efficiency instead of many. Section 7.1 explains the different techniques with critique and Section 7.2 implements one of the techniques on most of the metrics mentioned above with some improvements.

7.1 Different Techniques to Unify the Different Units of GPIs

There are so many metrics with different units that measure different aspects in the whole IT system. This makes it very hard to distinguish between a data center and another based upon their energy efficiency or "greenness". Thus unifying the different units of the green metrics or having one value indicating the "greenness" of a data center will allow for evaluating energy consumption in a comparative manner. There are two main approaches that can be taken to unify the units of the different GPIs based on recent works found in [1, 15]. The first is aggregating the values for the GPIs via methods such as normalization, and the second is determining one metric or value that is enough to indicate the total "greenness" of a data center. Next we will describe and compare three techniques based upon the two approaches given above.

The first technique described in [1] uses an aggregating approach to unify the units. It based on the idea of entropy (E_S) and its associated threshold (T_E). The entropy is an indicator that measures the compliance of the level of greenness of a service center with specific energy-saving requirements [25]. This technique uses entropy to describe the "greenness" of a data center. The technique simply measures if for a given context situation meaning part of a system being considered, the evaluated entropy is below a predefined threshold (T_E), then all the GPIs/KPIs for this context are fulfilled, otherwise some GPIs/KPIs did not meet the expectations and "greener" techniques need to be adopted to bring down the entropy level. The way to calculate the entropy that they defined sets a predefined threshold or a related policy for each GPI/KPI and evaluates whether the threshold is satisfied or not. The entropy is given by this equation: $E_s = \sum f_n$ where the formula f_n is defined as $f_n : pn\{0, 1\}$ and indicates whether the policy p_n associated to a GPI/KPI condition is fulfilled or not.

The second technique described also in [1] uses an aggregating approach to unify the units as well. In this technique a new metric name Green Level describes the energy efficiency or "greenness" of a data center. The value of this metric can be evaluated as the sum of the weights of each metric multiplied by a function that produces a dimensionless value normalized between 0 and 1 for all the metrics. The Green Level is given by this equation:

$$GREEN\_LEVEL = \sum w_n * f_n(GPI_n)$$

where the weights w_n are defined based on the preferences given by the involved stakeholders and the fn function is the evaluation function that allows to compare heterogeneous indexes to be compared [1].

The third technique described in [15] uses one metric that is enough to indicate the total "greenness" of a data center. The authors in [15] use a thermodynamic metric, *exergy* that measures the available energy in a system in joules to quantify the environmental impact from the operational phase of the system lifecycle. Exergy is formally defined as the maximum amount of useful work that can be derived relative to the surrounding reference state. For servers, the exergy consumption can be approximated by the total electricity consumption during operation. The authors state that when electricity is consumed by a server, it is converted to heat and loses most of the potential for useful work. To evaluate the server electricity consumption, for each component the maximum power rating is used to model how the power varies with workload and time periods of operation. In [15], the exergy consumption in infrastructures is also considered since cooling and power delivery infrastructure account for a large fraction of the total electricity consumption. The exergy consumption is modelled using the power usage effectiveness (PUE) metric. PUE is given by PUE = (operational power + infrastructure)/operational power.

The above techniques contain some vague areas that can certainly raise some questions and critiques. In the first scheme in which entropy is used to aggregate the different values of the different metrics, only a rough estimate is given to evaluate the total "greenness" of a data center. The measurement (entropy) is computed based on if a threshold value is fulfilled or not rather than the actual measurement value of each metric. The second technique aggregates all the values of the metrics after normalizing them to a range between {0–1} and multiplying by their weights. This technique gives a much more accurate measurement as compared to the first technique. The last technique which determines the total greenness of a data center based on exergy consumption only focuses one issue which is energy wastage/power consumption instead of taking into account the rest of the measurements computed by the other metrics. This value is not applicable in all cases; either a large data center or a small data center. Thus this technique does not reflect the overall picture because not all the metrics are considered. Overall, we believe the most appropriate technique is the second one.

7.2 Implementation of the Second Technique

The technique we will be implementing on one of the above metrics for all the classes is the second one based on normalizing the metric values to a range between {0–1}, multiplying by their weights, and then aggregating the results.

The above frameworks are structured based on the relationships between the metrics found in the same class. Similarities and replacements found between metrics in each class may also indicate correlations between the class. The idea of correlations between the metrics found in each class help us a lot in decreasing the total number of metrics to unify only since they measure the amount of similarities between them. It is not enough to define similarities between the metrics without a correlation measurement. Correlations may indicate direct proportionality between the metrics of one section or more than one section in a specific class as represented in the above frameworks. For example an increase in the PUE value indicates a greater amount of energy going to waste which as a result increases as well the Data Center Temperature metric. These two metrics as a result are correlated. The amount of correlation found between a metric and another with different units can be calculated as follow and found in [31]:

$$\text{Correlation} = \frac{\int_0^T F_i(t)G_j(t)dt}{\sqrt{\int_0^T F_i^2(t)dt \int_0^T G_j^2(t)dt}}$$

where F is value of metric number i, G is the value of metric number j, and T is the time over which the metrics are measured. We assume here that the metric values change with time. If the data is given in discrete time, then the above integrations should be replaced by summations. We note that the correlation lies between 0 and 1. The smaller the correlation value, the more independent are the two metrics are. The correlations found between the metrics will decrease the number of metrics leading to a less number of units and thus ease the process of unification. We reduce the number of metrics in a class based on strong/high correlations between them. If a number of metrics with the same unit have correlations between them are strong then we may exclude some metrics, but if the correlations are weak then they should be all kept.

Now we will give an example of strong and weak correlations between four different metrics (these are hypothetical measurements):

$$F_i = \text{CPU Usage} = 0.9, 0.95, 0.8, 0.85, 0.90, 0.95$$

$$G_j = \text{Userver} = 0.7, 0.75, 0.6, 0.7, 0.75, 0.85$$

$$\text{Correlation} = (.9 * .7 + .95 * .75 + .8 * .6 + .85 * .7 + .9 * .75 + .95 * .85)/$$

$$\sqrt{(.7^2 + .75^2 + .6^2 + .7^2 + .75^2 + .85^2)(.9^2 + .95^2 + .8^2 + .85^2 + .9^2 + .95^2)}$$

$$= 0.99$$

0.99 indicates a strong correlation between the CPU Usage and Userver metrics.

$$F_i = \text{CPU Usage} = 0.15, 0.35, 0.45, 0.55, 0.65, 0.75$$

$$G_j = \text{Unetwork} = 0.9, 0.8, 0.7, 0.6, 0.4, 0.1$$

$$\text{Correlation} = (.15 * .9 + .35 * .8 + .45 * .7 + .55 * .6 + .65 * .4 + .75 * .1)/$$

$$\sqrt{(.15^2 + .35^2 + .45^2 + .55^2 + .65^2 + .75^2)(.9^2 + .8^2 + .7^2 + .6^2 + .4^2 + .1^2)}$$

$$= 0.67$$

0.67 indicates a weak correlation between the CPU Usage and the Userver metrics.

After finding the correlations which reduce the number of metrics we are working with, we can apply the second technique presented in [1] based on normalizing the metric values to a range between {0–1}, multiplying by their weights, and then aggregating the results. After obtaining the correlations between the metrics of the same class for the four classes, we will have a smaller applicable number of metrics which we can convert to a unified unit: cost instead of Green Level, the metric used in [1]. The equation that represents the "greenness" of a data center in terms of cost is given by:

$$\text{Cost} = \sum_{i=1}^{n} w_i f_i(M_i)$$

where i is from 1 to n (the total number of metrics).

This equation is inspired by the work found in [1] except that is measures the greenness of a data center is cost instead of its green level. There is no need in explaining each variable in the equation because it is exactly the equation presented in [1] and explained in the section above. M_i is the actual

Table 8 Metrics where the expected cost is low and $F_i(M_i) = M_i$.

Metrics	Range	$F_i(M_i)=$
CPU Usage, Memory Usage, I/O Device Usage, Storage Usage, DH-UR (server), DH-UR (storage), Storage Utilization, Ustorage, Unetwork, ITEU, SI-POM, H-POM, CO2, CUE, WUE, UDC, RHD, Data Center Temperature, SHI, AEU	0-1	M_i

Table 9 Metrics where the expected cost is low and $F_i(M_i) = 1 - 1/M_i$.

Metrics	Range	$F_i(M_i)=$
PUE, pPUE, BTU, Human Resources, Infrastructural Costs, Carbon Credit, Consumables Index, Total Cost of Ownership	1-∞	$1-1/M_i$

value of each metric and the $f_i(M_i)$ function conditions each metric to a dimensionless value that lies between 0 and 1 and thus makes the different units transparent. The closer this value is to 0 the less is the cost and vice versa. Next, we define the value $f_i(M_i)$ for all the metrics above is given below:

If the value in terms of cost for the metric should be *high* to indicate a more "green" data center and:

- If the range of M_i is between (0–1), with 1 being the most favourable value
    Then the value $f_i(M_i) = 1 - M_i$.
- If the range of M_i is between (1–∞), with ∞ being the most favourable value
    Then $(1/M_i)$ will give a range (1–0).

If the value in terms of cost for the metric should be *low* to indicate a more "green" data center then:

- If the range of M_i is between (0–1), with 0 being the most favourable value
    Then the value $f_i(M_i) = M_i$.
- If the range of M_i is between (1–∞), with 1 being the most favourable value
    Then $(1/M_i)$ will give a range (1–0) and then a suitable definition is
    $f_i(M_i) = 1 - 1/M_i$ giving a range (0–1).

In Tables 8, 9, 10, and 11 the metrics are divided based on the range of their actual values (M_i) and their associated normalized value. Note that metrics with a non-dimensionless metric should be divided by the maximum

Table 10 Metrics where the expected cost is high and $F_i(M_i) = 1 - M_i$.

Metrics	Range	$F_i(M_i)=$
DH-UE, Userver, Server Compute Efficiency, DCD, SWaP, DCcE, DCiE, DCPE, COP, CPE, ERE, GEC, DPPE, FE, VAC, RCI, RHI, CI, CSE, CSS	0-1	1- M_i

Table 11 Metrics where the expected cost is high and $F_i(M_i) = 1/M_i$.

Metrics	Range	$F_i(M_i)=$
ITEE	0-1	1/M_i

possible their value to normalize them and allow them to fall between the range {0–1}.

We give now an example that compares between two different applications that perform the same function and are run in the same data center. The weights being used in this example are hypothetical.

Metric	System A	System B
CPU Usage	95%	83%
Memory Usage	93%	60%
I/O Usage	40%	75%

In this case, $M = 3$, weight$_1$ for CPU usage is 0.6, weight$_2$ for Memory usage is 0.6, weight$_3$ for I/O usage is 0.9.

$$\text{Cost}_A = \sum_1^3 (.95 * .6) + (.93 * .6) + (.4 * .9) = 1.49$$

$$\text{Cost}_B = \sum_1^3 (.83 * .6) + (.6 * .6) + (.75 * .9) = 1.53$$

As is shown, System A is better because its cost is less due to the difference in the weight values between the metrics.

8 Conclusion

Green Performance Indicators play a key role in building more energy efficient data centers. Although many metrics have been defined to measure all the factors that waste energy, there is no standard or popular framework till now that is to used define the "greenness" of a data center. We collected most of the GPIs defined by different associations and categorize them into four main classes accepted by the EU Project GAMES in which each metric is defined and compared with other metrics in the same class and frameworks are structured for each class. We also showed how GPIs contribute to every phase in the green computing process. Since GPIs measure different factors related to energy consumption, they may have different units. If unified into one single generic unit, it will be simpler to compare between the GPIs and indicate where energy is mostly wasted, and allow for the comparison between different IT systems or applications that do the same function. We compare and critique between three different techniques that define approaches for unification, and implement a technique on the metrics collected with improvements. In the future the complex correlations and dependences between GPIs in the same class can be clarified by applying experiences based on the collection of historical monitoring data for the corresponding GPIs.

References

[1] A. Kipp, T. Jiang, M. Fugini, and I. Salomie. Layered green performance indicators. Future Generation Computer System, Elsevier, 28(2):478–489, 2012.

[2] S. Tiwari. Need of green computing measures for Indian IT industry. Journal of Energy Technologies and Policy, 1:18–24, 2011.

[3] M. Jamalzadeh and N. Behravan. An exhaustive framework or better data centers' energy efficiency and greenness by using metrics. Indian Journal Computer Science and Engineering, 2:813–821, 2012.

[4] L. Wang and S.U. Khan. Review of performance metrics for green data centers: A taxonomy study. The Journal of Supercomputing, 1007–1024, 2011.

[5] D. Chen, E. Henis, C. Cappiello, A. Mello, T. Jiang, J. Liu, and A. Kipp. Usage centric green performance indicators. ACM SIGMETRICS Performance Evaluation Review, 39:92–96, 2011.

[6] A. Kipp, T. Jiang, and M. Fugini. Green metrics for energy-aware IT systems. In Proceedings of International Conference on Complex Intelligent and Software Intensive Systems, pp. 241–248, 2011.

[7] M. Fugini and J.A.P. Maestre. Energy analysis of services through green metrics: Towards green certificates. Information Systems Technology and Management, 285:247–258, 2012.

[8] I. Goiri, J.Ll. Berral, J. Fito, F. Julia, R. Nou, J. Guitart, R. Gavalda, and J. Torres. Energy-efficient and multifaceted resource management for profit-driven virtualized data centers. Future Generation Computer System, 28:718–731, 2012.

[9] B. Andersen and T. Fagerhaug. Green performance measurement. International Journal of Business Performance Management, 1:171–185, 1999.

[10] M. Talebi and T. Way. Methods, metrics, and motivation for a green computer science program. ACM SIGCSE Bulletin, 41:362–366, 2009.

[11] J. Willians and L. Curtis. Green: The new computing coat of arms. In IEEE Computer Society IT Pro, pp. 12–16, 2008.

[12] J. Pierson. Green task allocation: Taking into account the ecological impact of task allocation in clusters and clouds. Journal of Green Engineering, 1(2):129–144, 2011.

[13] B. Goska, J. Postman, M. Erez, and P. Chiang. Hardware/software codesign for energy-efficient parallel computing. In Proceedings of International Conference for High Performance Computing Networking Storage and Analysis, ACM, 2011.

[14] F. Rahman, C. O'Brien, S. Ahamed, H. Zhang, and L. Lui. Design and implementation of an open framework for ubiquitous carbon footprint calculator applications. In Sustainable Computing: Informatics and Systems, 1(4):257–274, 2011.

[15] J. Chang, J. Meza, and P. Ranganathan. Totally green: Evaluating, and designing servers for lifecycle environmental impact. In ASPLOS, ACM, pp. 25–36, 2012.

[16] R. Larrick and K. Cameron. Consumption-based metrics: From autos to IT. IEEE Computer Society, 97–99, 2011.

[17] L. Hilty and W. Lohmann. The five most neglected issues in green IT. Council of European Professional Informatics Societies CEPIS, XII:11–15, 2011.

[18] G. Sissa. Utility computing: Green opportunities and risks. Council of European Professional Informatics Societies CEPIS, XII:16–21, 2011.

[19] J. Taina. Good bad and beautiful software – In search of green software quality factors. Council of European Professional Informatics Societies CEPIS, XII:22–27, 2011.

[20] P. Lago, T. Jansen, and M. Jansen. The service greenery-integrating sustainability in service oriented software. Software Research and Climate Change, 2010.

[21] Z. Jun. Research on greenness evaluation index system of enterprise information system. International Symposium, pp. 209–212, 2011.

[22] C. Hsu, J.A. Kuehn, and S.W. Poole. Towards efficient supercomputing: Searching for the right efficiency metric. ICPE ACM, pp. 157–162, 2012.

[23] http://en.wikipedia.org/wiki/Greencomputing.

[24] M. Uddin and A. Abdul Rahman. Energy efficiency and low carbon enabler green IT framework for data centers considering green metrics. Renewable and Sustainable Energy Review, 16(6):4078–4094, 2012.

[25] T. Cioara, I. Anghel, I. Salomie, G. Copil, D. Moldovan, and B. Pernici. A context aware self-adapting algorithm for managing the energy efficiency of it service centers. The Ubiquitous Computing Journal (UBICC), 2010.

[26] Green Grid, http://www.the greengrid.org/, 1 November 2010.

[27] The Uptime Institute: www.uptimeinstitute.org, 1 November 2010.

[28] Nomura Research Institute: http://www.nri.co.jp/english/.

[29] Emerson Corporation: http://www.emerson.com/en-US/Pages/Default.aspx.

[30] Green IT Promotion Council. http://www.greenit-pc.jp/e/.

[31] B.P. Lathi and Z. Ding. Modern Digital and Analog Communication Systems, Oxford, 4th Edition, Section 2.5, New York, 2010.

Biographies

Sara S. Mahmoud received her B.Sc. degree in Computer Engineering from Kuwait University, Kuwait, in 2010. Currently, she is pursuing her M.Sc. in Computer Engineering with the Department of Computer Engineering at Kuwait University and also working as a teaching assistant in the same department. Her research interests include green computing and computer networks.

Imtiaz Ahmad received his B.Sc. in Electrical Engineering from University of Engineering and Technology, Lahore, Pakistan, an M.Sc. in Electrical Engineering from King Fahd University of Petroleum and Minerals, Dhahran, Saudi Arabia, and a Ph.D. in Computer Engineering from Syracuse University, Syracuse, New York, in 1984, 1988 and 1992, respectively. Since September 1992, he has been with the Department of Electrical and Computer Engineering at Kuwait University, Kuwait, where he is currently a professor. His research interests include design automation of digital systems, high-level synthesis, and parallel and distributed computing.

Energy Efficient Wireless Communications through Cooperative Relaying

M. Pejanovic-Djurisic, E. Kocan and M. Ilic-Delibasic

Faculty of Electrical Engineering, University of Montenegro,
Podgorica, Montenegro; e-mail: {milica, enisk, majai}@ac.me

Received 31 August 2012; Accepted: 27 September 2012

Abstract

It has been shown that cooperative communication schemes can solve many of the issues faced by future broadband WWAN and WLAN networks. It is about communication concept based on resource sharing and coordination among terminals in wireless network that provides significant performance improvements in terms of increased coverage, data rates, capacity, reliability, spectral and energy efficiency. This paper gives detailed overview of cooperative communication concept based on replacement of direct communication link between source and destination with several shorter links using network terminals called relays. Several, so called fixed relaying techniques, are described: amplify-and-forward fixed gain (AF FG), amplify-and-forward variable gain (AF VG) and decode-and-forward (DF). Appropriate analytical models for outage probability, bit error rate and system capacity values are presented. Further on, assuming Rayleigh fading channels, comparison of the presented relaying techniques are performed, enabling identification of optimal signal transmission scenarios for cooperative communication systems.

Keywords: cooperative relaying, amplify-and-forward, decode-and-forward, outage probability, BER, capacity, energy efficiency.

Journal of Green Engineering, Vol. 3, 71–90.

1 Introduction

With the increasing demand for new smart services and applications, a significant focus is on the further development of wireless communication networks, so that the required high throughputs and energy efficiency will be provided. However, it is well known that wireless signal transmission imposes serious challenges in fulfilling those demands, due to the complex nature of wireless radio channel. Thus, in defining the adequate technical solutions for future broadband wireless networks, all relevant characteristics of this specific transmission medium have to be taken into account. That is why research efforts have been directed towards new solutions and techniques that would support high data rates and higher capacities of future wireless systems, with the better coverage and energy efficiency at the same time.

Cooperative communication concept based on resource sharing and co-ordination among units of wireless network, where one or more intermediate nodes (relays) intervene in the communication between a transmitter and a receiver can solve many of the issues faced by future broadband WWAN and WLAN networks. The idea of exploiting benefits of diversity systems by mutual cooperation among terminals originates from 1970s [1]. It is about a concept that attains broader coverage by splitting the communication link from the source to the destination into several shorter links/hops. Since future wireless communications are likely to take place at higher and less congested frequency bands, where path loss is larger, and higher transmission powers are needed to keep the same coverage area, such concept is likely to be adopted in next generation systems. One of the main advantages of this communication technique is that it distributes the use of power throughout the hops, reducing the need to use a large power at the transmitter, which results in extended battery life and lower level of interference introduced to the rest of the network. Moreover, different energy aware schemes can be used to further save energy in transmitting data from the relays to the destination, such as different cooperative algorithms, power allocation, relay selection, sharing or distributing tasks among cooperating entities, etc.

This paper gives detailed explanations of relay aided communication concept. In its simplest form, a relay based system has just one relay station (R) and the entire communication process between a source of information (S) and a destination terminal (D) is performed over R [2]. This represents *dual-hop* relay system with three communication terminals (Figure 1), where R receives a signal from the source, performs adequate processing and after that transmits it towards the destination. In order to achieve full advantages

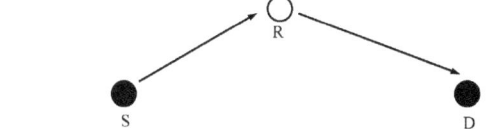

Figure 1 Dual-hop relay system.

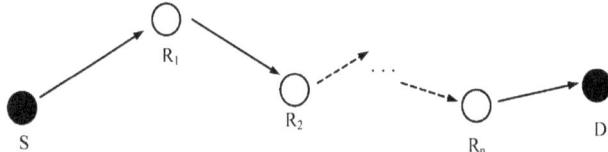

Figure 2 Multi-hop relay system.

of the relay implementation, it is necessary to obtain that the communication channel between S and R is orthogonal with the communication channel between R and D. The required orthogonality can be realized in the frequency domain, in the time domain, or using signals which are orthogonal in space-time constellation.

A relay system with three communication terminals represents the simplest example of relay aided communication network and it could be considered as a particular case of a multi-hop relay system. It is clear that an extended wireless link, covering greater distances between S and D, cannot always be successfully realized including just one R. If n denotes the number of relay stations participating in communication between S and D, then the multi-hop relay system is characterized with the S-D communication link being divided into $n + 1$ links (hops). There, each relay station communicates with the two neighboring terminals, as it is illustrated in Figure 2.

The above mentioned dual-hop and multi-hop cooperative relay systems are basically introduced in order to better cope with the effects of severe propagation losses present in wireless communications over longer distances. At the same time, their implementation contributes towards overall capacity improvements of wireless systems, enabling extension of their coverage range by maintaining the message transmission in the areas where it would not be possible without relay stations. However, relaying concept could also be implemented in a form of diversity system, transmitting multiple signal replicas towards destination terminal. In a simple three terminals scenario, diversity

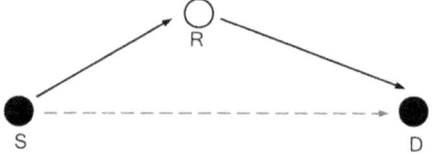

Figure 3   Dual-hop relay system with diversity.

is actually formed if additional direct communication link between S and D exist (Figure 3).

When this scenario with three communication terminals is considered, assuming that the orthogonality between S-R and R-D links is achieved in the time domain, i.e. that R operates in half-duplex mode, it is possible to identify different models for the realization of diversity transmission. Actually, as the communication process between S and D is divided in two time intervals, or two phases depending on terminals which participate in a particular phase, several models of this cooperative diversity could be recognized. Dual-hop relay system with diversity

Following growing interest for MIMO (Multiple Input Multiple Output) systems in wireless communications, additional focus has been directed towards relaying after presenting the idea of creating virtual MIMO system using single antenna relay terminals [3]. MIMO systems, already incorporated in different wireless network standards, offer significant performance improvements of wireless systems characterized with the communication channel exposed to fading and other known impairments. However, a practical implementation of this concept might be a problem in certain conditions due to limitations related with placing multiple antennas on a single terminal. That is why virtual, or distributed, MIMO system has emerged as an interesting solution for obtaining benefits of MIMO concept in a scenario with single antenna terminals (Figure 4).

Another option for incorporating relay systems in wireless environments can be created with wireless mesh networks [4], which include mesh clients, mesh nodes and gateways (Figure 5). In this configuration mesh nodes actually represent relay stations that can communicate with all neighboring terminals (nodes). Thus, the existence of such redundant communication links makes mesh networks highly reliable.

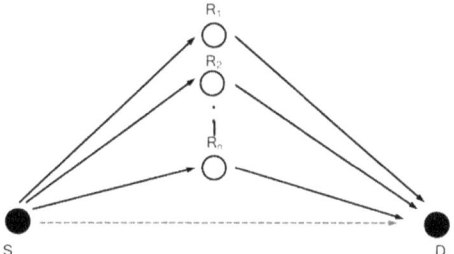

Figure 4   Virtual (distributed) MIMO.

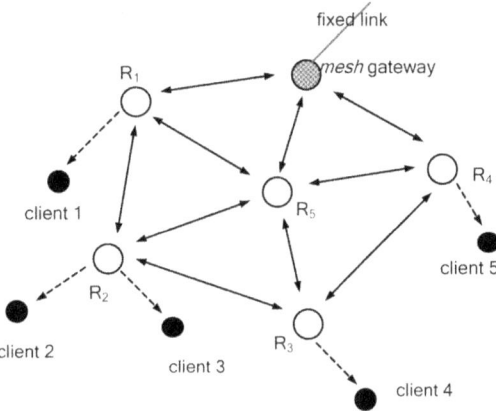

Figure 5   Mesh network.

2 Relaying Techniques

Performances of relaying systems highly depend on signal-to-noise ratio (SNR) of particular communication links, as well as the implemented signal processing method at the relay. With regard to algorithms of signal processing and forwarding applied at relay stations, relaying techniques can be classified as [5]:

- Transparent relaying techniques that perform simple power scaling and/or phase rotation, i.e. linear transformation of a signal received at R,
- Regenerative relaying techniques that include modifications of a signal waveform.

Transparent relaying techniques are Amplify-and-Forward (AF), Linear-Process-and-Forward, Nonlinear-Process-and-Forward. AF attracts most of the attention in relay systems considerations, where R receives a signal from a source, amplifies it and then forwards it towards a destination. In contrast with this approach, Decode-and-Forward (DF) is typical example of regenerative relaying technique. Thus, a relay station with DF first fully decodes a received signal, re-encodes it and then retransmits it towards a destination. Perform-ances of the mentioned AF and DF relaying techniques highly depend on signal-to-noise ratio (SNR) of particular communication links, what can be considered as a limitation factor for identification of a generally optimal re-laying technique. That is why a choice of an optimal relaying technique could be done only for a well defined specific communication scenario. When com-paring the two relaying techniques AF and DF, considered as the most used ones, it can be noticed that AF is characterized with the simpler realization and less delay introduced at relay stations. On the other side, it has a signi-ficant disadvantage in the fact that amplifying a signal it amplifies a present noise as well. DF relaying has specific advantage as it allows completely separated optimizations of S-R and R-D links, due to the fact that the process of re-encoding at R could be done with a code which is the most adequate for R-D link no matter what was a code used for signal transmission over S-R link.

Taking into account the importance of AF and DF relaying techniques that can be considered as bases for all the other signal processing and forwarding techniques used in relay aided communications, it is necessary to describe their behavior and performances in detail.

3 Amplify and Forward Relay Technique

As has already been mentioned, the Amplify-and-Forward (AF) relay tech-nique represents one of the two basic methods used for processing a signal received at the relay station. Depending on the way the signal is amplified, the following types of AF systems can be recognized:

- AF with fixed gain (FG),
- AF with variable gain (VG).

In AF FG relaying, R amplifies the received signal always with the same level, no matter the actual conditions on the S-R link. On the other hand, in the AF VG system the relay station permanently estimates the S-R link and, depending on the channel state information, determines the level of signal scaling applied.

For the elementary configuration of a dual-hop relay system with three communication terminals, shown in Figure 1, the signal received at R is given as:

$$y_R(t) = x(t)h_1(t) + n_1(t) \tag{1}$$

with $x(t)$ being a data symbol emitted by the source at the time instant t, $h_1(t)$ is the fading amplitude of the S-R channel and $n_1(t)$ is an additive white Gaussian noise, with variance N_{01}. The signal received at D depends on the way the signal is amplified.

3.1 AF with Fixed Gain

In AF relay systems with fixed gain G, the signal received at the destination can be represented with:

$$y_D(t) = Gy_R(t)h_2(t) + n_2(t) = Gx(t)h_1(t)h_2(t) + Gn_1(t)h_2(t) + n_2(t), \tag{2}$$

where $h_2(t)$ is the fading amplitude of the R-D channel at the given instant of time and $n_2(t)$ is an additive white Gaussian noise with variance N_{02}. The above relation illustrates the following two important characteristics of AF FG systems: (1) if fixed gain G is neglected, the total fading amplitude at the time instant t introduced over the S-R-D channel can be obtained by multiplication of fading amplitudes on S-R and R-D links at the same instant of time, i.e. $h(t) = h_1(t) \cdot h_2(t)$, and (2) these systems are characterized with the cumulative propagation of noise from S to the destination. Usually, in the systems with the fixed gain, G is taken to be:

$$G = \sqrt{\frac{\varepsilon_R}{\mathbf{E}[|y_R(t)|^2]}} = \sqrt{\frac{\varepsilon_R}{\varepsilon_S \mathbf{E}[|h_1(t)|^2] + N_{01}}} \tag{3}$$

In (3), ε_R and ε_S denote energy of the symbols emitted by R and S, respectively, and $\mathbf{E}[\cdot]$ is expectation operator. AF relay system which implementing this type of gain at relay station is usually called semi-blind AF relay system, or AF relay system with the average power limitation. There, it is assumed that R has information on the S-R channel statistics, i.e. on the average fading power, which is assumed to have relatively slow variations. Therefore, there is no need for continual estimation of the S-R channel. Using for the analyzed AF relay system with fixed gain, the following expression for the instantaneous signal-to-noise ratio at D can be written [6]:

$$\gamma_{end} = \frac{\gamma_{SR}\gamma_{RD}}{\frac{\varepsilon_R}{G^2 N_{01}} + \gamma_{RD}}, \tag{4}$$

where

$$\gamma_{SR} = \frac{\varepsilon_S}{|h_1(t)|^2} N_{01} \quad \text{and} \quad \gamma_{RD} = \frac{\varepsilon_R}{|h_2(t)|^2} N_{02} \tag{5}$$

denote the instantaneous signal-to-noise ratios of the S-R and R-D links, respectively. Instantaneous SNR at the system receiving end is given with:

$$\gamma_{end} = \frac{\gamma_{SR}\gamma_{RD}}{1 + \bar{\gamma}_{SR} + \gamma_{RD}} \tag{6}$$

$\bar{\gamma}_{SR}$ represents the average SNR of the S-R link.

Despite the fact that AF relay systems with fixed gain have come into research focus long after AF relay systems with variable gain, their performances have already been analyzed in different communication scenarios, as well as for various types of communication channels, [6–8].

When performance evaluation of wireless communication systems is considered, the outage probability is often used as a relevant parameter. Conventionally, it describes probability that SNR on a link falls below predetermined threshold value, γ_{th}. The assumed system, where the communicating terminals transmit on orthogonal channels, is often denoted as noise-limited system. In other words, it means that the interference caused by other nodes is below the noise level. In such a system an outage is usually caused by deep fades that drive SNR below γ_{th}. When AF relaying techniques are concerned, outage probability is declared as probability that instantaneous SNR at the system receiving end, γ_{end}, falls below γ_{th}, i.e.:

$$P_{out} = P_r[\gamma_{end} < \gamma_{th}] \tag{7}$$

For dual-hop AF FG relay system, when S-R and R-D channels have Rayleigh narrowband fading statistics, the outage probability is derived in [6] as:

$$P_{out} = 1 - 2\sqrt{\frac{\rho\gamma_{th}}{\bar{\gamma}_{SR}\bar{\gamma}_{RD}}} e^{-\gamma_{th}/\bar{\gamma}_{SR}} K_1\left(2\sqrt{\frac{\rho\gamma_{th}}{\bar{\gamma}_{SR}\bar{\gamma}_{RD}}}\right) \tag{8}$$

where $K_1(\cdot)$ represents the first order, modified Bessel function of the second kind, and the coefficient ρ is equal to:

$$\rho = \frac{G^2\varepsilon_R}{N_{01}}. \tag{9}$$

$\bar{\gamma}_{SR}$ and $\bar{\gamma}_{RD}$ are average SNRs on the S-R and R-D links, respectively.

Probability density function (PDF) and the moment generating function (MGF) of the end-to-end SNR, for the assumed dual-hop AF FG relay system in Rayleigh fading environment, are also given in [6] The PDF of SNR is:

$$f_{\gamma,\text{end}}(\gamma) = \frac{2e^{-(\gamma/\bar{\gamma}_{SR})}}{\bar{\gamma}_{SR}} \left[\sqrt{\frac{\rho\gamma}{\bar{\gamma}_{SR}\bar{\gamma}_{RD}}} K_1 \left(2\sqrt{\frac{\rho\gamma}{\bar{\gamma}_{SR}\bar{\gamma}_{RD}}} \right) \right.$$
$$\left. + \frac{\rho}{\bar{\gamma}_{RD}} K_0 \left(2\sqrt{\frac{\rho\gamma}{\bar{\gamma}_{SR}\bar{\gamma}_{RD}}} \right) \right], \tag{10}$$

with $K_0(\cdot)$ denoting the zero order modified Bessel function of the second kind.

Knowing PDF, MGF of the end-to-end SNR for dual-hop AF relay system with fixed gain is derived as:

$$\mathcal{M}_{\gamma,\text{end}}(s) = \frac{1}{(\bar{\gamma}_{SR}s + 1)} + \frac{\rho\bar{\gamma}_{SR}s \exp\left[\frac{\rho}{\bar{\gamma}_{RD}(\bar{\gamma}_{SR}s+1)}\right]}{\bar{\gamma}_{RD}(\bar{\gamma}_{SR}s + 1)^2} E_1 \left(\frac{\rho}{\bar{\gamma}_{RD}(\bar{\gamma}_{SR}s + 1)} \right) \tag{11}$$

In the above relation $E_1(\cdot)$ denotes the exponential integral function.

The BER performance and capacity of AF FG relay systems can be analyzed using PDF and MGF of the received SNR. The upper bound of the average ergodic capacity for AF relay system with FG, and Rayleigh fading channel on both hops, can be defined as [8]:

$$C = \frac{1}{2}\mathbf{E}(\log_2(1 + \gamma_{\text{end}})) \le \frac{1}{2}\log_2(1 + \mathbf{E}(\gamma_{\text{end}})) \tag{12}$$

where a multiplication with 1/2 is introduced as a consequence of the fact that communication process is realized in two time intervals. Such definition of the average ergodic capacity represents the system capacity normalized over the unit bandwidth, for the channel which is considered as ergodic. Expectation of SNR at the system receiving end is:

$$\mathbf{E}(\gamma_{\text{end}}) = \bar{\gamma}_{SR}e^{\frac{\theta_R}{2\bar{\gamma}_{RD}}} \left[2W_{-2,1/2} \left(\frac{\theta_R}{\bar{\gamma}_{RD}} \right) + \sqrt{\frac{\theta_R}{\bar{\gamma}_{RD}}} W_{-3/2,0} \left(\frac{\theta_R}{\bar{\gamma}_{RD}} \right) \right] \tag{13}$$

where $W_{k,\mu}(z)$ denotes the Whittaker function, and coefficient θ_R is equal to:

$$\theta_R = \frac{\varepsilon_R}{G^2 N_{01}}. \tag{14}$$

3.2 AF with Variable Gain

Using Eq. (1) which describes the signal received at R, the signal at D of the AF relay system with variable gain can be represented with:

$$y_D(t) = G(t)y_R(t)h_2(t) + n_2(t)$$
$$= G(t)x(t)h_1(t)h_2(t) + G(t)n_1(t)h_2(t) + n_2(t). \qquad (15)$$

As can be noticed, the applied gain is a function of time, having variations which follow changes of the S-R channel in accordance with:

$$G(t) = \sqrt{\frac{\varepsilon_R}{\varepsilon_S |h_1(t)|^2 + N_{01}}} \qquad (16)$$

With R with variable gain, it becomes possible to compensate deleterious effects related with the signal propagation over S-R link, so that a relay station R always transmits the signal with the same power. This is the reason why AF relay system with this type of gain is also known as AF system with the instantaneous power limitation. It is clear that the AF VG system is more complex than the AF FG, as it requires permanent estimation of the S-R channel. Introducing the gain factor $G(t)$ into (15), the fading amplitude of the whole S-R-D channel at the given time t is obtained as

$$h(t) = \frac{\sqrt{\varepsilon_R} h_1(t)h_2(t)}{\sqrt{\varepsilon_S |h_1(t)|^2 + N_{01}}}. \qquad (17)$$

The above given relation shows that end-to-end characteristics of uplink and downlink channels are not identical. Following the expression for the signal received at the destination (15), instantaneous SNR at the receiving end of the system with variable gain can be defined with

$$\gamma_{end} = \frac{\gamma_{SR}\gamma_{RD}}{1 + \gamma_{SR} + \gamma_{RD}}. \qquad (18)$$

Analyses of the AF VG relay system, focused on its outage probability, capacity as well as on its bit error rate, have been performed for different communication scenarios including dual-hop, multi-hop or cooperative diversity configurations [9–11]. It has been shown that, even for the elementary dual-hop configuration, derivation of the closed form relation for the PDF of the received SNR, in AF VG relay system with Rayleigh narrowband fading statistics, might be very complex without certain approximations. That is why performance analyses of these systems usually assume that a relay station introduces variable gain $G(t)$ given as [9]

$$G(t) = \sqrt{\frac{\varepsilon_R}{\varepsilon_S |h_1(t)|^2}}. \tag{19}$$

The above given expression gives a relation for the instantaneous received SNR which is much more suitable for further mathematical manipulations:

$$\gamma_{\text{end}} = \frac{\gamma_{SR}\gamma_{RD}}{\gamma_{SR} + \gamma_{RD}}. \tag{20}$$

The probability that the above defined γ_{end} falls below the predefined threshold γ_{th} is given as [9]

$$P_{\text{out}} = 1 - 2\sqrt{\frac{\gamma_{\text{th}}}{\bar{\gamma}_{SR}\bar{\gamma}_{RD}}} \exp\left[-\gamma_{\text{th}}\left(\frac{1}{\bar{\gamma}_{SR}} + \frac{1}{\bar{\gamma}_{RD}}\right)\right] K_1\left(2\sqrt{\frac{\gamma_{\text{th}}}{\bar{\gamma}_{SR}\bar{\gamma}_{RD}}}\right). \tag{21}$$

The PDF of the SNR at the receiving end, in the assumed dual-hop scenario with Rayleigh fading statistics on each particular channel (hop), can be determined as [9]

$$f_{\gamma,\text{end}}(\gamma) = \frac{2\gamma \exp\left[-\left(\frac{\gamma}{\bar{\gamma}_{SR}} + \frac{\gamma}{\bar{\gamma}_{RD}}\right)\right]}{\bar{\gamma}_{SR}\bar{\gamma}_{RD}}$$

$$\times \left[\frac{(\bar{\gamma}_{SR} + \bar{\gamma}_{RD})}{\sqrt{\bar{\gamma}_{SR}\bar{\gamma}_{RD}}} K_1\left(\frac{2\gamma}{\sqrt{\bar{\gamma}_{SR}\bar{\gamma}_{RD}}}\right) + 2K_0\left(\frac{2\gamma}{\sqrt{\bar{\gamma}_{SR}\bar{\gamma}_{RD}}}\right)\right]. \tag{22}$$

When the MGF of the received SNR is considered, for the assumed communication scenario and the case when $\bar{\gamma}_{SR} = \bar{\gamma}_{RD} = \bar{\gamma}$, it is given with [9]:

$$M_{\gamma,\text{end}}(s) = \frac{\sqrt{\frac{\bar{\gamma}}{4}s\left(\frac{\bar{\gamma}}{4}s + 1\right)} + \text{arcsinh}\left(\sqrt{\frac{\bar{\gamma}}{4}}\right)}{2\sqrt{\frac{\bar{\gamma}}{4}s}\left(\frac{\bar{\gamma}}{4}s + 1\right)^{3/2}} \tag{23}$$

Using the PDF of the received SNR given with (22) and (12), an upper bound of the average ergodic capacity of AF VG relay system, for Rayleigh fading statistics on both hops, is defined as [8]:

$$E(\gamma_{\text{end}}) = \frac{4\sqrt{\pi}\beta_R^2}{3.3233(\varepsilon_R + \beta_R)^3}\left[{}_2F_1\left(3, \frac{1}{2}; \frac{7}{2}; \frac{\varepsilon_R - \beta_R}{\varepsilon_R + \beta_R}\right)\right.$$

$$\left. + \frac{3\varepsilon_R}{\varepsilon_R + \beta_R}{}_2F_1\left(4, \frac{1}{2}; \frac{7}{2}; \frac{\varepsilon_R - \beta_R}{\varepsilon_R + \beta_R}\right)\right]. \tag{24}$$

In the above relation, coefficients ε_R and β_R are equal to:

$$\varepsilon_R = \frac{1}{\bar{\gamma}_{SR}} + \frac{1}{\bar{\gamma}_{RD}} \quad \text{and} \quad \beta_R = \frac{2}{\sqrt{\bar{\gamma}_{SR}\bar{\gamma}_{RD}}} \tag{25}$$

while $_2F_1(\cdot, \cdot; \cdot; \cdot)$ is a Gaussian hypergeometric function.

4 Decode and Forward Relay Technique

The Decode and forward (DF) technique in dual-hop relay communication system is performed over two completely separated subchannels, since R first decodes the signal received from S and then the signal is re-encoded at R and transmitted to the destination. If the signal received at R is represented as in (1), then the signal received at the destination becomes:

$$y_D(t) = \hat{x}(t)h_2(t) + n_2(t), \tag{26}$$

where $\hat{x}(t)$ denotes an estimation of the signal $x(t)$, obtained at R. The decoding process which is applied at R introduces evident system perform-ance improvements since the total noise at the destination is decreased when compared with AF relay systems. At the same time, it becomes possible to implement modulation schemes at S-R and R-D links which are not necessar-ily identical, so that optimal modulations can be applied in accordance with SNR levels at particular links. Thus, communication process is divided in two asymmetric time intervals, where the longer time interval is always dedicated to the communication process over the link with smaller SNR. This presents another advantage of the system with DF relaying, when compared with AF systems, and it is clear that it leads towards better BER performance. On the other side, DF signal processing at R can be at the origin of certain drawbacks in the case of channels with severe fading. When BER is concerned, degrad-ation appears if there is an error in the decoding process engaged at the relay station, since erroneously decoded symbols are then further transmitted to D. There is no doubt that characteristics of R-D link can also contribute to overall BER performance degradation of DF relay system, as additional errors might be introduced in the signal recuperation at terminal D. In order to reduce those negative implications of the error propagation and to improve BER performance, different encoding schemes for error detection and correction can be applied in the process of signal regeneration at DF relay stations [12].

In DF systems, an outage event occurs if either one of the links is in outage, i.e. if SNR in either of them falls below γ_{th}. In dual-hop DF relay systems, it is the complement event of having both links operating above predefined γ_{th}. Hence, for a Rayleigh fading scenario on both links, it is equal to:

$$P_{out} = 1 - \left(\int_{\gamma_{th}}^{\infty} \frac{1}{\bar{\gamma}_{SR}} e^{-(\gamma/\bar{\gamma}_{SR})} d\gamma \right) \left(\int_{\gamma_{th}}^{\infty} \frac{1}{\bar{\gamma}_{RD}} e^{-(\gamma/\bar{\gamma}_{RD})} d\gamma \right)$$

$$= 1 - e^{-\gamma_{th}(1/\bar{\gamma}_{SR}+1/\bar{\gamma}_{RD})} \tag{27}$$

When the achievable capacity of DF dual-hop relay system is concerned, it is of the uttermost importance to notice that it is limited with the characteristics of the worse of the two links engaged in the communication process. Namely, ergodic capacity of DF dual-hop relay system can not be higher than the ergodic capacity of the link (S-R or R-D) which has the lower instantaneous signal-to-noise ratio, i.e.:

$$C = \frac{1}{2} \min\{\log_2(1 + \gamma_{SR}), \log_2(1 + \gamma_{RD})\}. \tag{28}$$

Knowing the ordered statistics of random variables, and assuming Rayleigh fading channel on both hops, capacity of dual-hop DF relay systems can be analyzed. In such a scenario, ergodic capacity for DF can be defined as

$$C = \frac{1}{2\ln(2)} \exp\left(\frac{1}{\bar{\gamma}_{SR}} + \frac{1}{\bar{\gamma}_{RD}} \right) E_i \left(\frac{1}{\bar{\gamma}_{SR}} + \frac{1}{\bar{\gamma}_{RD}} \right). \tag{29}$$

5 Performance Analysis of Relay Systems

5.1 Outage Probability

Outage probability as a performance measure shows the probability that link quality does not satisfy the required level. Thus, it might be useful to perform comparison of outages probabilities of the analyzed relaying techniques with the case of direct transmission. In this way, an insight if the assumed AF and DF relay systems may improve the quality of the equivalent link between the S and D can be achieved. For the sake of attaining fair comparison conditions, it is assumed that the total transmitted powers, P_T, in the case of direct transmission and in all the concerned dual-hop relay systems, are the same. Moreover, we took equal power allocation among the S and the R station, i.e. $P_S = P_R = P_T/2$. Now, the average SNRs on S-R and R-D links can be written as $\bar{\gamma}_{SR} = A_1 P_S$ and $\bar{\gamma}_{RD} = A_2 P_R$, respectively, where A_1 and A_2 include parameters as antenna gains, path loss, noise power and similar. For example, if using Friis propagation model, A_i, $i = 1, 2$, can be written in the form:

$$A_i = \frac{G_{t,i} G_{r,j} \lambda^2}{(4\pi)^2 d_i^\alpha L N_{0,j}}, \tag{30}$$

where $G_{t,i}$ is the transmitter antenna gain on the i-th hop, $G_{r,i}$ is the receiver antenna gain, λ is the wavelength, d_i is the distance between the transmitter and receiver on the i-th hop, L is the system loss factor, $\alpha = 2$ for free space and $3 < \alpha < 4$ in urban environment, while $N_{0,i}$ is the noise variance at the i-th hop. Without loss of the generality, we took that the transmitter antenna gains at S and R are equal, $G_{t,1} = G_{t,2}$, and the receiver antenna gains at R and D are also equal, $G_{r,1} = G_{r,2}$, as well as that the noise variances at R and D are the same, $N_{0,1} = N_{0,2}$. Moreover, we assumed that in the case of relayed transmission, S, R and D are placed on a straight line, and that all the links are affected by the same shadowing environment. The average SNR at D in the case of direct transmission can be written as $\bar{\gamma}_{SD} = A_{eq} P_T$, where for this simplified propagation model, by taking $\alpha = 3$, A_{eq} is related to A_1 and A_2 through:

$$A_{eq} = \frac{A_2}{(1 + (A_2/A_1)^{1/3})^3}. \tag{31}$$

Outage probability for the case of direct transmission in Rayleigh fading environment is equal to:

$$P_{out} = \int_0^{\gamma_{th}} \frac{1}{\bar{\gamma}_{SD}} e^{-\gamma/\bar{\gamma}_{SD}} d\gamma = 1 - e^{-\gamma_{th}/\bar{\gamma}_{SD}}. \tag{32}$$

Figure 6 shows the outage probability of the considered relaying techniques, as well as of the direct transmission case, as a function of the total transmitted power in the communication systems, P_T. It is taken that $A_1 = 2$ and $A_2 = 10$, while $\gamma_{th} = 0$ dB. The advantage of using relayed transmission over the direct transmission is evident, regardless of the relaying technique implemented. For the presented outage probability values, the total transmitted power saving is between 2 and 2.5 dB when using relaying techniques, in comparison with direct transmission. The achieved power saving for the same link level quality implies, in its turn, less introduced interference to other communicating nodes in comparison to direct transmission. This is additional benefit of cooperative communication concept. Further performance improvements in terms of outage probability in relay systems may be attained through using optimal power allocation strategies over the hops, for a given power budget [14].

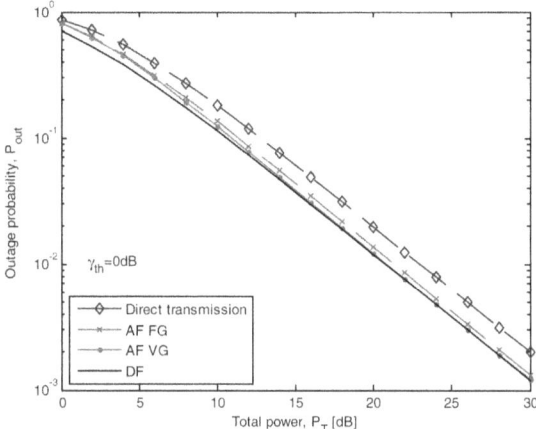

Figure 6 Outage probability as a function of total transmitted power in the system, P_T.

If we compare outage probability performances of the considered relaying strategies, it can be seen from Figure 6 that DF relaying technique have the best performance for all P_T values. However, the difference in outage probability among the three considered relaying techniques is very small. Thus, for example, for the P_T above 15 dB, AF VG relay system has completely the same outage probability performance as the DF relay system.

5.2 BER Performance

BER of the wireless communication system can be determined using the known MGF of SNR at the system receiving end [13]. Thus, for example if a DPSK (Differential Phase Shift Keying) modulation is applied, the bit error rate is defined by

$$P_b = 0.5 \mathcal{M}_{\gamma,\text{end}}(1). \tag{33}$$

Introducing (11), or (23) into (33), BER expressions for dual hop AF FG and AF VG relay system are obtained, in that both channels are characterized with Rayleigh fading statistics.

In DF relay systems a signal is transmitted over two cascade links and its decoding is done twice. If the transmission implies a binary signal with two possible symbol states (DPSK or BPSK), an error will appear at the final destination terminal only if an error in the signal detection is performed once

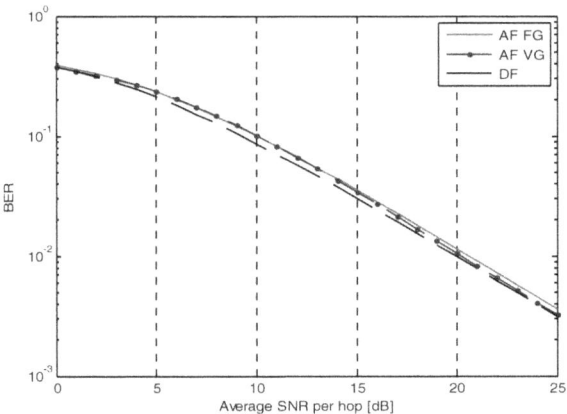

Figure 7 BER performance comparison for AF and DF dual-hop relay systems.

(either on the first or on the second link):

$$P_b = 1 - [(1 - P_{b1})(1 - P_{b2}) + P_{b1} P_{b2}] = P_{b1} + P_{b2} - 2P_{b1} P_{b2} \quad (34)$$

where P_{b1} and P_{b2} are bit error rates at the first and the second link (hop), respectively. When DPSK is assumed, probability of error for each of the links is given with and, for Rayleigh fading statistic on both links, the overall BER is obtained in the form:

$$P_b = \frac{1 + \bar{\gamma}_{SR} + \bar{\gamma}_{RD}}{2(1 + \bar{\gamma}_{SR})(1 + \bar{\gamma}_{RD})}. \quad (35)$$

Figure 7 illustrates BER graphs for DPSK modulated DF, AF FG and AF VG dual-hop relay systems operating in the assumed scenario with Rayleigh narrowband fading statistics on S-R and R-D links. It is assumed that the average SNR of the S-R link is equal to the average SNR of the R-D link, i.e. $\bar{\gamma}_{SR} = \bar{\gamma}_{RD}$.

The graphs presented in Figure 7 give interesting and in a certain manner surprising results, since differences among BERs for the three systems considered are unexpectedly small, having in mind considerable differences related with the complexity of the systems and their implementations. Still, it can be seen that DF relay technique has the best BER performance. However, even when compared with AF FG system which has the worst BER performance, SNR gain does not overpass 0.5 dB for the whole range of BER values analyzed. At the same time, for higher SNRs per hop (over 24 dB), BER results for AF VG relay system are identical with the ones obtained for DF relay technique.

In addition, Hasna and Alouini [14] considered the case when the links are highly unbalanced in terms of their average fading power. In that case, optimal power allocation enhances the system performance, in terms of BER and outage probability. Interestingly, they also show that nonregenerative systems with optimum power allocation can outperform regenerative systems with no power optimization, i.e. same performance can be obtained with less power.

5.3 Capacity Performance

In order to gain a complete insight into benefits and trade-offs related with the choice of a particular signal processing/forwarding technique applied at relay stations, it is necessary to take into account achievable capacity as well. Figure 8 shows appropriate graphs for ergodic capacity of AF FG, AF VG and DF relay systems. Presented results are obtained by simulation under assumption that S-R and R-D links are characterized with Rayleigh narrowband fading statistic with the average SNR at the first hop being equal to the average SNR at the second hop.

It can be clearly noticed that AF VG relay system achieves the lowest ergodic capacity for the whole range of SNRs per hop, excluding very small SNRs (up to 2.5 dB) where its capacity performance is slightly better when compared with AF system with fixed gain. DF relay system has the highest ergodic capacity for SNRs per hop below 12.5 dB, while for higher SNRs it is AF FG system which has the best values of ergodic capacity. Thus, for example, the presented graphs show that for ergodic capacity being equal to 3 b/s/Hz, AF FG relay system has SNR gain of almost 1dB in comparison with DF relay system, while its SNR gain in comparison with AF VG relay system is a bit less than 2.5 dB.

6 Conclusions

This paper gives detailed explanations of relay aided communication concept, with the description of techniques applied for the message processing and/or forwarding at the relay nodes. Assuming a wireless channel with Rayleigh fading, comparisons of the presented relaying techniques with direct transmission case in terms of outage probability is performed, proving the benefits of energy saving achieved through relayed transmission. Moreover, mutual

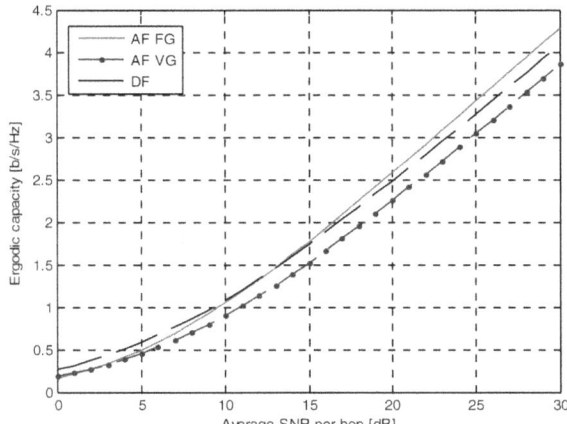

Figure 8 Comparison of ergodic capacities of dual-hop relay systems.

performance comparison in terms of BER and ergodic capacity of the analyzed relaying systems is presented, enabling identification of optimal signal transmission scenarios for cooperative communication systems.

Following the description of individual relaying techniques, as well as the comparison of theirs outage probability, BER and capacity performances presented in this paper, it is quite clear that it is not possible to identify a technique which would be absolutely superior in terms of performances for the whole range of SNR values. However, depending on characteristics of S-R and R-D links and on a performance being of interest for a specific communication process, there is always a possibility to assume which of the three analyzed systems will provide the best transmission quality and reliability.

In addition, one of the main advantages of such system is that it distributes the use of power throughout the hops. This implies more energy efficient system that will provide longer battery life and lower interference introduced to the rest of the network. Furthermore, it enables power allocation strategy that can further enhance the system performance and energy efficiency.

References

[1] E.C. van der Meulen. Three-terminal communication channels. Advanced Applied Probability, 3, 120–154, 1971.
[2] F.H.P. Fitzek and M.D. Katz (Eds.). Cooperation in Wireless Networks. Springer, 2006.
[3] A. Nosratinia and A. Hedayat, Cooperative communications in wireless networks. IEEE Comm. Mag., 74–80, October 2004.

[4] G.R. Hiertz, D. Denteneer, S. Max, R. Taori, J. Cardona, L. Berlemann, and B. Walke. IEEE802.11s: The WLAN mesh standard. IEEE Wireless Comm., 17(1), 104–111, February 2010.

[5] J.N. Laneman, D.N.C. Tse, and G.W. Wornell. Cooperative diversity in wireless networks: Effcient protocols and outage behavior. IEEE Trans. Inform. Theory, 50, 3062–3080, December 2004.

[6] M.O. Hasna and M.S. Alouini. A performance study of dual-hop transmissions with fixed gain relays. IEEE Trans. Wireless Commun., 3, 1963–1968, November 2004.

[7] G.K. Karagiannidis. Performance bounds of multihop wireless communications with blind relays over generalized fading channels. IEEE Trans. Wireless Commun., 5, 498–503, March 2006.

[8] G. Farhadi and N.C. Beaulieu. On the ergodic capacity of wireless relaying systems over Rayleigh fading channels. IEEE Trans. on Wireless Comm., 7(11), 4462–4467, November 2008.

[9] M.O. Hasna and M.S. Alouini. End-to-end performance of transmission systems with relays over Rayleigh-fading channels. IEEE Trans. Wir. Comm., 2(6), 1126–1131, November 2003.

[10] A. Ribeiro, X. Cai, and G.B. Giannakis. Symbol error probability for general cooperative links, IEEE Trans. Wireless Comm., 4(3), 1264–1273, May 2005.

[11] H.A. Suraweera, R. Louie, Y. Li, G.K. Karagiannidis, and B. Vucetic. Two hop amplify-and-forward transmission in mixed Rayleigh and Rician fading channels. IEEE Trans. Comm., 13(4), April 2009.

[12] T. Wang, A. Cano, G.B. Giannakis, anad J.N. Laneman. High-performance cooperative demodulation with decode-and-forward relays. IEEE Trans. on Comm., 55(7), July 2007.

[13] M.K. Simon and M.-S. Alouini. Digital Communication over Fading Channels, 2nd ed. Wiley, New York, 2005.

[14] M.O. Hasna and M.S. Alouini. Optimal power allocation for relayed transmissions over Rayleigh-fading channels. IEEE Trans. Wireless Comm., 3(6), 1999–2004, November 2004.

Biographies

Milica Pejanovic-Djurisic is Full Professor in Telecommunications at the University of Montenegro, Faculty of Electrical Engineering, Podgorica, Montenegro. Professor Pejanovic-Djurisic graduated in 1982 from University of Montenegro with BSc degree in Electrical Engineering. She received her MSc and PhD degrees in Telecommunications from University of Belgrade. For a period of two years, Professor Pejanovic-Djurisic also performed research in mobile communications at University of Birmingham, UK. She has been teaching at University of Montenegro telecommunications courses on graduate and postgraduate levels, being the author of four books, many strategic studies, and participating in numerous internationally funded research teams and projects. She has published more than 200 scientific papers in international and domestic journals and conference proceedings. Professor Pejanovic-Djurisic has organized several workshops and given tutorials and speeches at many scientific and technical conferences. Her main research interests are: wireless communications theory, wireless networks performance improvement, broadband transmission techniques, optimization of telecommunication development policy. She has considerable industry and operating experiences working as industry consultant and Telecom Montenegro Chairman of the Board. Professor Pejanovic-Djurisic has also been involved in activities related with telecommunication regulation. Being an ITU expert, she participates in a number of missions and ITU workshops related with regulation issues, development strategies and technical solutions.

Enis Kocan is a teaching/research assistant at the University of Montenegro, Faculty of Electrical Engineering, Podgorica, Montenegro. He received the BSc and MSc degrees in electronics engineering from the University of Montenegro, in 2003 and 2005, respectively. He defended his Ph.D. thesis at the same University in 2011, in the area of mobile communications, with the topic being OFDM-based cooperative communications for future generation mobile cellular systems. His major research interests are in digital communications over fading channels, physical layer aspects of wideband cooperative systems and multi-hop communications. Dr. Kocan has published and presented more than 40 scientific papers in international and national scientific journals, international and regional conferences and is co-author of the book *OFDM-Based Relay Systems for Future Wireless Communications*, published by River Publishers, Denmark in 2012.

Maja Ilic-Delibasic is a teaching/research assistant at the University of Montenegro, Faculty of Electrical Engineering, Podgorica, Montenegro. She received the BSc and MSc degrees in electronics engineering from the University of Montenegro, in 2003 and 2006, respectively, and is currently working toward her Ph.D. degree at the Center for Telecommunications, Faculty of Electrical Engineering, University of Montenegro. Her main research interests are: wireless communications theory, wireless networks performance improvement, physical layer aspects of wideband cooperative systems.

TVWS Radio Spectrum Utilization: Use Case of India-Looking Forward

Tanuja Satish Dhope (Shendkar)[1], Dina Simunic[1]
and Ramjee Prasad[2]

[1] Faculty of Electrical Engineering and Computing, University of Zagreb, Croatia;
e-mail: tanuja_dhope@yahoo.com, dina.simunic@fer.hr
[2] Center for TeleInfrastruktur, Aalborg University, 9220 Aalborg, Denmark;
e-mail: prasad@es.aau.dk

Received 31 August 2012; Accepted: 27 September 2012

Abstract

The complexity of wireless networks requires careful design with special attention to energy and bandwidth efficiency. The energy efficiency is of more importance due to increasing penetration of wireless systems in different battery-operated applications as well as more conscious global view on the need for "Greening the Earth". Bandwidth efficiency is very important parameter, because it relates to frequency spectrum, which is naturally scarce resource. Thus spectrum sensing is an important part of "Green Engineering". The "Cognitive Radio" (CR) technology sheds new light on unavailability of spectrum by managing radio resources in more systematic and efficient way. Around the Globe there is a coordinated move for Digital Switch over (DSO) by discontinuing analogue television transmission. This "Digital Dividend" (DD) has created a new spectrum opportunities for many new wireless technologies. We focus on the scope and nature of opportunities for white space created by DD for Indian scenarios. We discuss use cases for the exploitation of Television White Space (TVWS) suitable for rural India based on user's and BS geo-location and user's mobility followed by QoS requirements and recent regulatory activities.

Journal of Green Engineering, Vol. 3, 91–112.

Keywords: DD, cognitive radio, IEEE 802.11 af, TV white spaces, rural India, use cases, regulatory aspects.

1 Introduction

Although spectrum is seen as a scarce natural resource, measurements show that often there are moments in time and space where the spectrum is not being fully utilized by the allocated services and therefore it is being used inefficiently [1].

ITU predicted that 1720 MHz spectrum will be required by the year 2020. Many new wireless services/applications cannot be rolled out due to non-availability of spectrum [2], which demands dynamic allocation of spectrum instead of static [1]. CR would help to meet the ever increasing demand of radio spectrum and help in managing resource in more systematic and in more efficient way. The basic idea of CR is to reuse the spectrum whenever it is vacant by the primary/licensed users (PUs). The secondary/unlicensed users (SUs) are required to perform the frequent spectrum sensing for detecting the presence of PUs [3]. TV switchover to full digital broadcast service (DD) creates new spectrum opportunities due to higher spectral efficiency compared to analogue services. This TVWS opens the door for CR technology.

Spectrum regulatory bodies in various countries are studying the pro and cons of CR devices. Some countries have already made provisions for CR. FCC has already made provision for use of CR device in TV bands [4]. Ofcom from UK has also studied CR in their spectrum framework review and made provision for use of unlicensed cognitive devices in TV band by using TVWS [5]. CEPT has implemented a CR based device to operate in TVWS [6]. The survey of spectrum utilization in India [7] is similar to other countries. The plot for spectrum usage at Mumbai and time [7] for a frequency band of 600–800 MHz is as shown in Figure 1. This plot of spectrum utilization for 600–800 MHz shows that the spectrum in these bands in India is underutilized and can be opened for CR application.

The organization of the paper is as follows: Section 2 describes the DD scenario. Section 3 deals with opportunities in TVWS, followed by use cases for TVWS usage in Section 4. Section 5 discusses QoS in TVWS cognitive access, followed by the regulatory aspects in TVWS in Section 6. Finally, in Section 7 we present the conclusion.

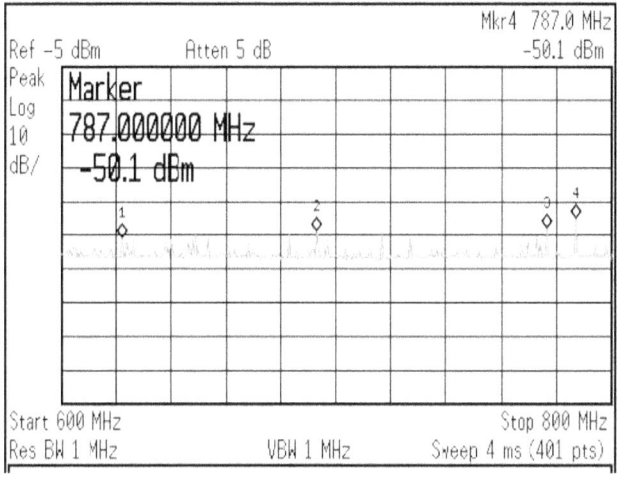

Figure 1 Frequency utilization in 600–800M Hz at one location (Mumbai) in India [7].

2 Digital Dividend

The DD scenario at international and in India is analyzed in this section.

2.1 International Scenario

The digital switchover process is underway but complete analogue switch-off is not easy. The full switch off of analogue services can have terrible consequences if viewers were not adequately prepared. So DD will require careful planning and the involvement of the entire broadcast industry.

The process of analogue switch-off will differ in countries depending upon the market configuration. Table 1 indicates an analogue switch-off (ASO) situation in various countries. The countries that have not done full ASO can take useful lessons from that have completed ASO about understanding best approaches, pitfalls that should be avoided, can help to ensure a successful process.

2.2 Indian Scenario

In India the Wireless Planning and Coordination (WPC) is the national radio regulatory authority responsible for frequency spectrum management. The National Frequency Allocation Plan (NFAP) forms the basis for development and manufacturing of wireless equipment and spectrum utilization. Accord-

Table 1 The ASO situation in various countries.

Country	Completion of ASO
UK, Luxemburg, the Netherlands, Finland, Sweden, Germany, Belgium, Denmark, Estonia, France, Czech Repulbic, Croatia, Switzerland, Malta, Slovenia, Japan, South Korea , Malta	Completed
Bulgaria, Cyprus, Greece	By the end of 2012
Australia, New Zealand, South Africa	2013
Pan-Arab	2014
Poland, Slovakia, India, Russia, Tunisia, Albenia, Cambodia	2015
Chile, Colombia	2017

Table 2 Spectrum allocation 470–806 MHz in India.

Band	Spectrum	Number of TV channels available in analogue mode/other services	TV channel number
UHF Band IV	470–582 MHz	14	21 to 34
		Mobile TV using DVB-H	26
UHF Band V	582–806 MHz	28	35 to 62
	610–806 MHz	BSNL, Defense operate point to point microwave links	–
	746–806 MHz	Public Protection and Disaster Relief (PPDR)	–
	> 806 MHz	Fixed, mobile services for transmission of data/voice, cellular mobile services	–

ing to NFAP-2011 [8], the vacant bands due to DD are allocated to other important services like International Mobile Telecommunications(IMT) services, data broadcasting, High definition TV(HDTV), ultra HDTV, mobile TV services, Super Hi-Vision (SHV) TV and Digital Terrestrial Television (DTT), etc. [9, 10].

In India, the 470–806 MHz band has been allotted to fixed, mobile broadcasting services on primary basis. In *UHF Band IV* (470–582 MHz) 14 TV channels, each of 8 MHz bandwidth, are available. Doordarshan is the only Government broadcaster that operates digital transmitters in four metros in this band. In *UHF Band V* (582–806 MHz) 28 TV channels each of 8 MHz bandwidth are available. Defence and BSNL are operating point-to-point microwave links in 610–806 MHz. Public Protection and Disaster Relief (PPDR) has some spots earmarked in 750–806 MHz. The UHF Band V above 806 MHz is also shared with other users of spectrum such as fixed and mobile services for transmission of data/voice and video (see Table 2). The complete switchover to digital transmission is a very challenging task in

India considering huge analogue TV sets in rural India and in more populated parts of India but it will be completed in 2015. This will be a slow process and hence Doordarshan suggested simulcast of analogue and digital transmission till complete switch off of analogue transmission [9].

3 Opportunities in TVWS

At present no specific regulations have been made for CR technology in India. CR technology should be implemented in those frequency bands where spectrum utilization is not high, the location of the base station is known and the receivers are robust against interference.

In this paper IEEE 802.22 and IEEE 802.11af standards are discussed which are based on CR operating in TVWS. The IEEE 802.22 standard is already finalized and in operation but the standardization process for IEEE 802.11af is not yet finished [11].

There are various opportunities for exploiting TVWS in an efficient manner. Some of them are discussed below.

3.1 Wide Area Coverage in Rural Areas (IEEE 802.22)

IEEE 802.22 wide regional area network (WRAN) aims to provide fixed wireless access with a typical cell radius of 33 km with Effective Isotropic Radiated Power (EIRP) of 4 W and a maximum of 100 km in rural and remote areas using CR technology in TVWS [12]. The spectral efficiency of 802.22 systems ranges from 0.5 to 5 bit/s/Hz. At an average value of 3 bit/s/Hz, the capacity of a 6 MHz channel can reach 18 Mbit/s. Broadband to rural areas might therefore prove to be very cheap.

3.2 Low Power Broadband: IEEE 802.11af

In some urban areas in e.g. Asia, Europe and USA there is not much white space (WS) for high power systems like fixed unlicensed devices that would operate from a fixed location and could be used to provide commercial services such as wireless broadband Internet access. However, there could be a potential for low power broadband systems like "portable/personal" unlicensed devices, such as wireless in-home local area networks (LANs) or Wi-Fi-like cards in laptop computers that exploit smaller portions of TVWS. 802.11af is a modified 802.11 standard, which operate in a range of TVWS using the properties of CR. The requirements specification of 802.11af system is formed, but the standardization process is not yet finished. It is also called as "White-Fi" or "Super Wi-Fi" ("White" due to work in unused TV band and "Super" due to its cognition properties).

Some promising applications of IEEE 802.11af with respect to Indian scenario are:

- *Remote monitoring of power plant*: nuclear power plants (e.g. Tarapore) or hydro power plants (Pophali) can be remotely monitored by smart grid technologies that exploits TVWS in order to improve system operation and control, manage its demand of electrical power, provide broadband Internet access to under-served areas (like tribal villages: Himachal Pradesh, Arunachal Pradesh and Mizoram, etc.) [13].
- *Backhaul for Wi-Fi* in campuses, businesses, hotels, theatres (Wi-Fi could be IEEE 802.11af or IEEE 802.11a/b/g/n) avoiding costly, time consuming and challenging cabling issues.
- *Traffic monitoring and precautionary measure for prevention from accident*: Traffic cameras can be installed at all major accident prone intersections to provide real-time traffic monitoring, reduce congestion, travel time and also fuel consumption.
- *Anti theft measure*: Theft like-car/personal/bicycle/motorcycle could be avoided by wireless camera connected to police for surveillance and also radios can be incorporated in city parks availing free Wi-Fi access to residents and city workers for surveillance.
- *Remote monitoring and managing heavy wetland areas* like the northeast states of India (Assam, Manipur, Sikkim, Mizoram and Nagaland) to comply with EPA regulations because these areas are hard to get, utilizing the WS network.
- *Medical monitoring, environmental monitoring, vehicular communications, cellular networks offloading and ad-hoc networks* (e.g. internal sensor network or mesh network communication).
- *Digital video broadcast for handheld (DVB-H) system with cognitive access*: DVB-H could be utilized as an unlicensed secondary network together with licensed network such as DVB-T, operating in UHF bands. Mobile TV providers prefer this band as it provides balance in the antenna size and coverage. It can be integrated with car terminals, portable digital TV sets and handheld portable convergence terminals. Still the big challenge with DVB-H by mobile operators is the issue of suitable business and revenue models. When implementing DVB-H based on CR, the following design considerations must be taken into account:

- Time slicing, in order to reduce the average power consumption of the terminal and to make possible smooth handovers. The time-slicing technique, enables considerable battery power-saving. Further, time-slicing allows soft handover if the receiver moves from network cell to network cell.
- An enhanced error protection scheme on the link layer to increase reception robustness for indoor and mobile contexts.
- In-depth symbol interleaver, for further improvement of the transmitted signal robustness in impulse noise conditions and mobile environment.

- *Public safety application with CR approach exploiting TVWS*: all public safety communication infrastructures in India use narrowband radios which limits them to 2-way voice communications with no inherent support for high-bandwidth transmission requirements such as interactive video communication, remote video surveillance of security or disaster sites and do not provide the level of secure communication required by India's security forces resulting in easy leak of information to rogue entities, e.g., terrorists. Although TETRA enhanced data services are available through WIMAX and LTE, but they are not sufficiently reliable as required by public safety measures. As Opportunistic spectrum allocations provide greater capacity for overloaded network and dynamic reconfiguration capability to better manage load and connectivity. The cognitive access to TVWS can be a viable alternative of a common band used for Public Safety. The CR spectrum broker can easily integrate a prioritizing mechanism that assigns TVWS for a specific disaster area based on spectrum availability information provided by the geo-location database.
- *Programme Making Special Events (PMSE)*: PMSE applications have already been already using TVWS as unlicensed users .As a result of DD, the amount of spectrum resources for PMSE were reduced. At the same time, there is an increasing demand for spectrum for PMSE [6]. The PMSE in TVWS employing CR will provide the improvement in terms of spectrum efficiency, reliability and productivity of existing applications.
  Typical applications of PMSE are, e.g., temporary connectivity to manage people assembled for exhibitions, festivals, election campaign, medical campaign, community awareness sessions and breaking news.

- *LTE extension (GSM/3G/HSPA)*: LTE has considerable flexibility, scalable channel bandwidths from 1.4 to 20 MHz which optimizes the use of radio spectrum by making use of new spectrum and re-farmed spectrum opportunities. It is best solution for deployment as a next generation cellular communication infrastructure. Operators evolving to LTE from GSM/WCDMA/HSPA will maintain full backward compatibility with legacy networks. The TVWS could also be exploited for managing peak traffic by obtaining and sharing channels on a temporary basis.

- *Femto-cell for wireless broadband in TVWS*: Femto-cells can be the next deployment choice for indoor environment as a complement to micro-cells for enhancing coverage and capacity in TVWS which can penetrate through walls and buildings, based on the concept of intra-operator spectrum WS reuse, multi-operator spectrum sharing and multi-service spectrum reuse exploiting the spectrum of multiple operators and of multiple non-cellular services (e.g. DTV broadcasts).

- *Rural broadband access-empowering rural India*: e-governance-agriculture-learning-healthcare and e-animal husbandry utilizing TVWS opportunities could benefit to the rural India by providing knowledge based economy, like in decreasing farmer suicides by boosting agriculture Gross Domestic Product (GDP), in decreasing primary school drop-out rate by providing video based trainings and in decreasing mortality rate viz. video conferencing for e-cardio testing and e-diagnosis facilities. Here we have given the example of e-animal husbandry application (see Figure 2). An online disease reporting system known as the "National Animal Disease Reporting (NADR)" system can be implemented using CR networks utilizing TVWS. With this system farmers will benefit by providing wireless connectivity among various units like Taluka unit, district unit, state unit, centre unit, veterinary (Vet.) dispensary, Taluka level NGO, etc. The Central Unit at Department of Animal Husbandry and Dairies (DAH&D) would be responsible for analysis and maintenance of data. All the disease related information would be available online, which will ensure transparency and benefit of all concerned.

Table 3 reflects the scenarios and typical ranges in IEEE 802.11 af standard utilizing TVWS based on above applications studied in Section 3.2.

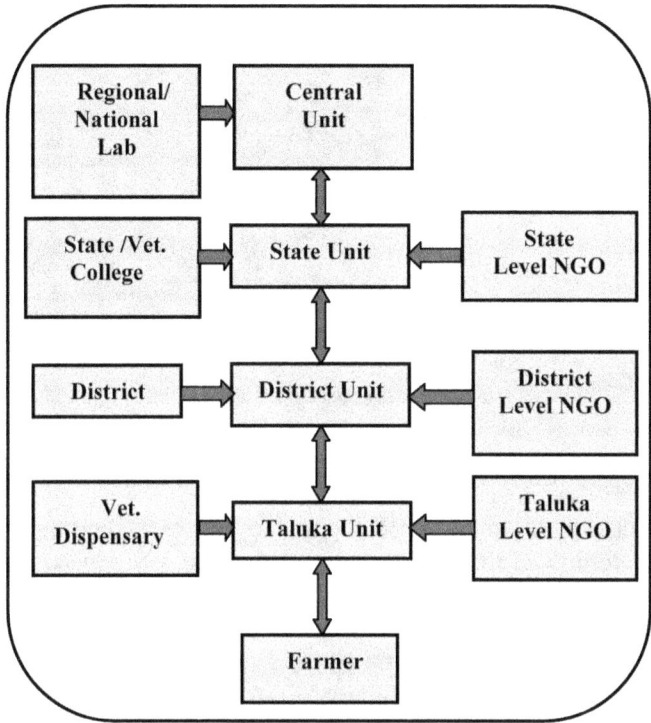

Figure 2 NADR System [15].

Table 3 Scenarios and typical ranges in IEEE 802.11 af standard based on CR networks utilizing TVWS.

Scenarios	Ranges
Backhaul	10 km
LTE extension in TVWS	0–10 km
Rural broadband	0–10 km
Terminal to terminal cellular	10–1000 m
Ad-hoc network in TVWS	0–100 m
Femto cell in TVWS	0–100 m

4 Use Cases for TVWS Usage

The ETSI RRS technical committee [13] is currently focusing on TVWS standardization, especially on both cellular and short range applications. These key use cases are implemented depending on user's and Base Station (BS) geo-location and user's mobility from rural India point of view.

Communication in
TVWS frequency bands

Figure 3 Mid/long range wireless access: no mobility (Ratnagiri, Maharashtra).

4.1 Use Case: Mid/Long Range Wireless Access

Internet access is provided by BS to the end user by exploiting TVWS over the ranges similar to the cellular system 0 to 10 km depending on user's mobility [15]. It is further classified as follows.

4.1.1 Mid/Long Range: No Mobility

Wireless access is provided from BS towards fixed mounted home access point/base station (see Figure 3). The geo-location for both BS and fixed device are well known.

4.1.2 Mid/Long Range: Low Mobility

In this scenario, wireless access is delivered from BS towards mobile devices where the user has low mobility, e.g. when walking or taking it around the farm. Sensing results for PUs fetched for the current location are not getting invalid because of its low mobility. GPS or cellular positioning system can be used for keeping track of mobile user. The geo-location from BS is well known.

4.1.3 Mid/Long Range: High Mobility

In this case, sensing results for PUs fetched from the current location are getting invalid because of fast mobility of the user, e.g. in a vehicle. The geo-location from BS is well known. More attention is needed for exploiting TVWS as it sets high constraints on PU sensing.

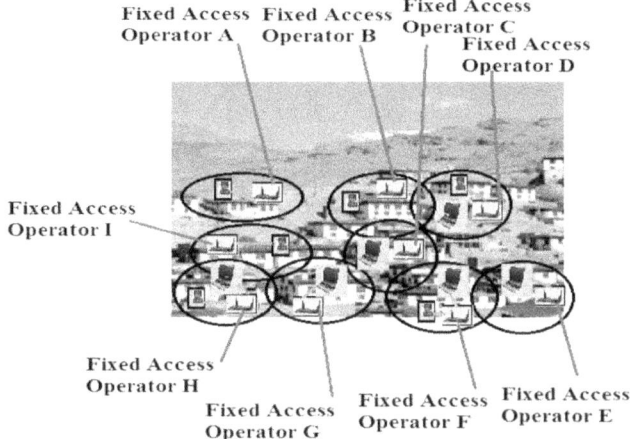

Figure 4 Uncoordinated networks, short range (villages in Maharashtra, India).

4.1.4 Centralized Network Management

In this case, we considered a network centric solution for allocating available TVWS for the user terminal to get connectivity. The available TVWS is considered based on location rather than on time. In rural India, TVWS would be largely available and in time. Once the terminal accesses the network, it can be left under the control of the network. For instance high layer signaling can be utilized for this purpose, e.g. handover command to hand-off to a new frequency or system broadcast messages can be used to notify terminals about change of the frequency.

4.2 Use Case: Short Range Wireless Access

Internet access is provided by BS to the end user by exploring TVWS over the ranges similar to the cellular system 0 to 50 m [15]. This use case is further studied in the next sections.

4.2.1 Uncoordinated Networks

In rural areas, the houses are large. All residents are equiped with their own Access Points (AP) operating in the WS frequency band. Thus, multiple coordinated networks can be an appropriate choice (see Figure 4). This scenario is also useful for surfing at the time of farming, diagnosis, in animal husbandry, etc.

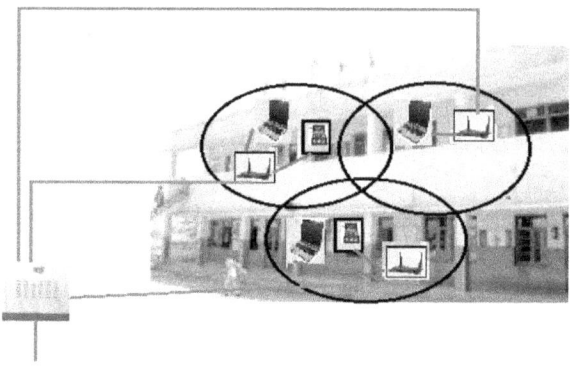

Fixed Access Operator

Figure 5 Coordinated networks, short range, and Fixed Access operator (schools, Gram panchayat or Tahasildar offices).

4.2.2 Coordinated Networks

WS networks in the close proximity are operated in a coordinated manner by the WS operator (see Figure 5). Examples of this kind of usage can be academic institute, the Gram Panchayat offices (i.e. local self-governments at the village which is responsible for overall village administration like looking after public health, education, keeping records of births/deaths, land details, etc.) or Tahasildar offices (i.e. head of Taluka administrations responsible for regulating the functionality of Gram Panchayat office, responsible for emergency services like natural calamity case such as lightening death, affected due to flood, etc., responsible for revenue collection such as rehabilitation/petitions/land reforms/enactment of various legislations and all other technical functions).

4.3 Hybrid of Uncoordinated and Coordinated Networks

Overall deployment can be thought of as a combination of both uncoordinated and coordinated networks. Such situation could be the case in government general hospital (primary health centre), in veterinary hospital and house for older people. In order to work properly, effective coexistence methods need to be in place for this scenario.

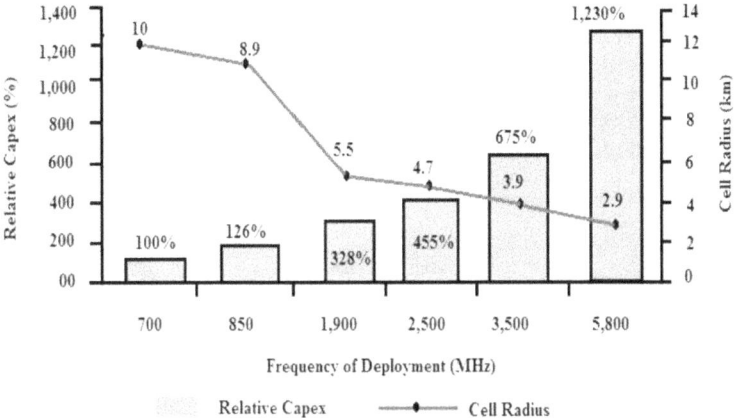

Figure 6 Relative capex vs. frequency of deployment vs. cell radius [17].

4.4 Use Case: Opportunistic Spectrum Access to TVWS by Cellular Systems in Absence of PU

In this case, TVWS slots are available sporadically for SUs such as for, e.g., multi-mode user terminals being able to operate as, among other systems, cellular systems in licensed and unlicensed spectrum. The supported unlicensed spectrum is assumed to include TVWS, i.e 470–806 MHz in India.

The cellular system in this band provides the following advantages:

- *Lighter infrastructure*: due to improved propagation characteristics in the TVWS bands compared to typical license band (GSM 1800 MHz, 2.1 GHz or 2.5 GHz bands for 3G/broadband wireless access,a larger cell size is chosen which will lead to lighter infrastructure and will result in reduced capital expenditure (CAPEX) (see Figure 6). Of course, in this case deployment in rural or high-cost regions becomes economically viable.
- Two or three times as many less sites required for initial coverage at 700 MHz compared to 2.1 or 2.5 GHz. An LTE network at 700 MHz would be 70% cheaper to deploy than an LTE network at 2.1 GHz-GSMA.

Selecting 700 MHz band for various application will saves the relative capital expenditure by two or three times as many less sites required for initial coverage at 700 MHz compared to 2.1 or 2.5 GHz (see Figure 7).

Table 4 Comparative study of cost vs propagation at frequency [17].

Cost	Propagation at frequency		
	700 MHz	1900 MHz	2400 MHz
Total Network cost @150K/cell	$150,000	$600,000	$1,500,000
Network Cost per Customer	$180	$725	$1820
# Mos. to Network Cost Break-even	9 months	36 months	91 months

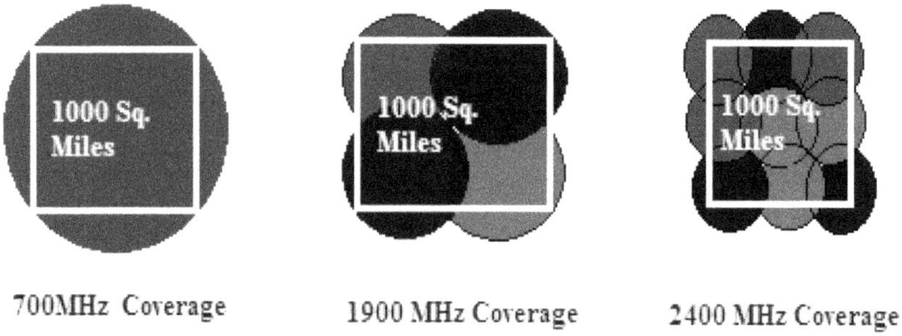

700MHz Coverage   1900 MHz Coverage   2400 MHz Coverage

Figure 7 Cell site coverage per thousand square miles (1000 sq. miles = 1609.34 km) [17].

- *Increased spectral efficiency* through reduced propagation loss (see Figure 8). A possible deployment choice is to keep a cell size as it is. The case of licensed band deployment leads to:
  1. Higher Quality-of-Service (QoS) achieved in a given cell. But these propagation characteristic may cause the interference issues which require an adequate handling like power management, suitable frequency re-use factor for TVWS.
  2. Higher QoS within a given cell at lower RBS/mobile devices (MDs) output power level. The inherent power consumption can be reduced.
  3. The hybrid solution of combination A and B is possible, i.e, a moderate reduction in RBS output power levels combined with a moderate improvement of the QoS.
  4. Increased spectral efficiency through extended macro diversity. A possible deployment choice is to keep cell size as it is the case of licensed band deployment. Then, joint operation of neighbouring RBS can be explored in order to achieve a higher macro-diversity gain in UL (multiple RBS are decoding jointly the received signals) or in the DL (multiple RBS are contributing to jointly optimized transmission).

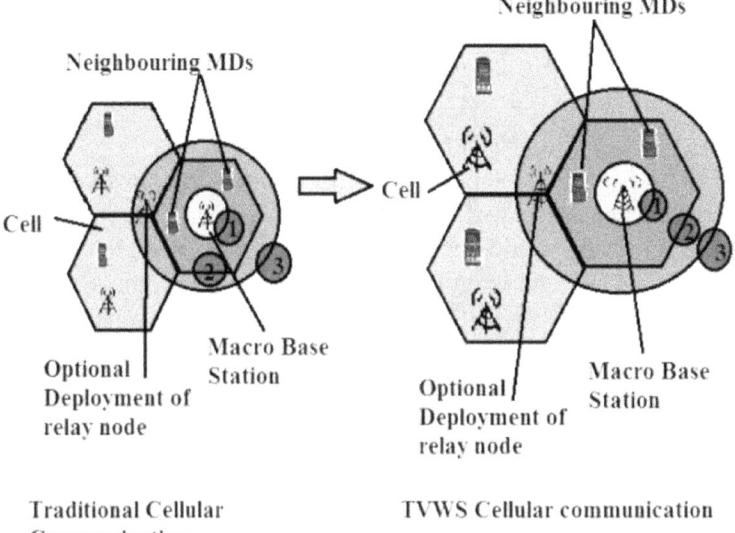

Figure 8 Improved coverage and reduced propagation loss in TVWS (note: symbols 1, 2 and 3 indicate decreasing throughput level, QoS, etc.).

4.5 Use Case: Ad-hoc Networking over White Space Frequency Bands

In this case, user devices and other devices like APs communicate with each other to share information or to run joint applications or services, or to execute other similar tasks. The communication happens by forming an ad-hoc network operating in WS frequency band. There can be two or more devices in the ad-hoc network formed like device-to-device connectivity (video communication in the case of NADR system) and ad-hoc networking (e-health; the devices can be operating a localized social networking service).

5 QoS in TVWS Cognitive Access

One of the challenging features of the TVWS is its variation across space and time. More specifically the available channels are not contiguous and vary from one location to another. In addition the white space available in a given location can vary with time if one or more of the TV band Pus start/stop operation. Opportunistic access to TVWS is interruptible in the sense that CR has to cease transmission immediately and relocate to a new band as soon as

Table 5 Technical characterstics for PMSE.

Applications	RF output	Max. RF Bandwidth (KHz)	Transmission height
PWMs	30 mw	200	1.5 to 12 m
Talkback	30 mw	20–50	1.5 to 12 m

the DVB-T or PMSE which is already using white space on secondary basis; appears. The delay associated with such relocations may face cognitive users with abrupt QoS degradation as communication peers need to coordinate the frequency transition, and many parameters across the protocol stack have to be reset to match the characteristics of the new frequency band. Therefore, cognitive radio links built on TVWS are inherently unreliable. The issue is on how to provide the best quality TVWS for secondary usage to maximize persistence of allocations while avoiding interference with primary users.

The QoS requirements for PMSE are given below.

5.1 High QoS

The wireless microphones and remote controls (e.g. fireworks) need a high reliable radio link interface else would cause degradation in quality of the production. The maximum tolerable delay 1min with required bandwidth of several of 100 KHz.

5.2 Moderate QoS

The equipment can have a certain probability of interference and can tolerate certain interference than mobile speech services. These devices are used for speech which is not intended for broadcasting. The required bandwidth of 20–50 KHz is sufficient with tolerable delay not more than 1 min.

6 Regulatory Activities Related to CR and TVWS

On the regulatory field, presently no specific regulation has been made for CR technology in India. In our views CR devices may be permitted to bands in which spectrum activity is low, location of base stations are known and receivers are robust against interference. Possible Indian approaches operating secondary TVWS need to be analyzed by national regulators WPC. Researchers must focus to the development of IEEE 802.11af and to the assessment of three cognitive techniques, like spectrum sensing [17–22], geo-location database and beacon, in order to provide protection to the licensed radio services.

Geo-location capability must be present in all fixed devices, with an accuracy of +/- 50 m. This position information is used to query a database for a list of available TV channels that can be used for cognitive devices operation. The database will include the information of all TV signals and may also have information on wireless microphone usage because of the challenging signal detection in low power wireless microphones. The following aspects have to be considered:

- Protection of digital broadcasting services including mobile TV.
- Protection of fixed and mobile services in the frequency bands of 470–520 and 520–585 MHz.
- Protection of services adjacent to a 470–806 MHz band like protection of mobile satellite services and broadcasting except aeronautical mobile satellite (R) service in the frequency band of 806–890 MHz and protection of IMT applications in the frequency band of 450.5-457.5 MHz paired with 460.5–467.5 MHz.
- Protection of PPDR communications.
- Definition of the requirements for the geo-location database approach.

As per [23–26], the regulation requirements for the IEEE 802.11af standard based on CR utilizing TVWS are as shown in Table 6. The exact regulatory framework of CR technology utilizing TVWS frequency band is yet to be finalized to guarantee an efficient exploitation of frequency bands in TVWS. As more research went along, it came clearer that the geo-location will play the main role of defining the channels that must be free of wireless sensing device transmissions. Hopefully the spectrum management and resource management of CR technology contain all the regulation constraints for all of the TVWS opportunities and that additional information will be taken from the network in order to exploit this portfolio to allocate the spectrum access using QoS based rules.

7 Conclusions

This paper has given the overview of DD scenarios and most promising applications for exploiting TVWS in Indian scenarios. The TVWS is expected to be available for the secondary spectrum access, especially in rural India. The applications like e-governance-education-agriculture-health-animal husbandry can take the benefits for improving Gross Domestic Product. Further,

Table 6 Regulation requirements for IEEE 802.11af.

Parameter	[23]	[24]	[25]	[26]
Quantized signal energy				
TV signals (8 MHz)	Not required	−114 dBm	Not required	Not required
Wireless microphone	Not required	−126 dBm in 200 kHz	Not required	Not required
Geo-Location accuracy	50 m	100 m	–	Not specified
Transmit power (fixed) EIRP	1 W (with max 6 dBi antenna gain)	100 mW	–	Local Specific
Transmit power (portable) EIRP	100 mW	100 mW	–	Local Specific
Transmit power in adjacent band to DTT signals	40 mW	20 mW	–	Local Specific
Out of band radiation	−55 dBm under the inband level	–	–	
Time between sensing	1 min	1 sec		

the use cases based on users and BS geo-location, as well as user's mobility, have been discussed based on Indian scenario showing that both cellular and short range applications can exploit TVWS. The QoS in CR exploiting TVWS has been discussed. Finally, assessment of regulatory framework is given to prevent harmful interference to licensed users. However, it does not provide insights to operate the available spectrum efficiently. Current research initiatives are tackling this objective.

References

[1] FCC Spectrum Policy Task Force Report. In Proceedings of the Federal Communications Commission (FCC'02), Washington, DC, USA, 2002.

[2] T. Dhope, D. Simunic, and M. Djurek. Application of DOA estimation algorithms in smart antenna systems. Studies in Informatics and Control, 19(4), 445–452, 2010.

[3] K-C. Chen and R. Prasad. Cognitive Radio Networks. John Wiley & Sons, 2009.

[4] FCC Spectrum Policy Task Force. http://www.fcc.org. ET Docket No. 04-186, 2008.

[5] Ofcom: Digital Dividend: Clearing the 800Mhz Band. http://www.ofcom.org.uk/consult/condocs/cognitive/.

[6] Report C from CEPT to the European Commission. http://www.erodocdb.dk/Docs/doc98/official/pdf/CEPTREP024.PDF.2008.

[7] P.S.M. Tripathi and Ashok Chandra. Radio spectrum monitoring for cognitive radio. In 2nd International Conference on WIRELESSVITAE, pp. 1–5, 2011.

[8] http://www.wpc.dot.gov.in/DocFiles/Draft_Channelling_Plan_for_NFAT-2011.

[9] www.wpc.dot.gov.in/DocFiles/Proposal_from_Doordarshan_to_JTG-India.doc.

[10] www.wpc.dot.gov.in/DocFiles/Proposal_fromCOAI.doc.

[11] 802.11 Working Group. IEEE P802.11af D0.06: Draft Standard for InformationTechnology. http//www.ieee802.org/11/. Accessed 2010.

[12] http://ieee802.org/22/.

[13] T. Dhope, D. Simunic, and R. Prasad. TVWS opportunities and regulatory aspects in India. In Proceedings of 14th International Symposium on WPMC'11, Brest, France, pp. 566–570, 2011.

[14] ETSI reconfigurable Radio Systems Technical Committee. Information available at http://portal.etsi.org.

[15] T. Dhope, D. Simunic, and R. Prasad. TVWS opportunities and regulation: Empowering rural India. In Proceedings of 14th International Symposium on WPMC'11, Brest, France, pp. 201–205, 2011.

[16] T.R. Dua. http://www.gisfi.org/wg_documents GISFI_SPCT_20101243.pdf, 2010.

[17] T. Dhope, D. Simunic, and R. Prasad. Hybrid detection method for cognitive radio. In Proceedings of 13th International Conference on SoftCOM, Split-Hvar-Dubrovnik, SS1-78741-1609, 2011.

[18] T. Dhope and D. Simunic. Performance analysis of covariance based detection in CR. In Proceedings 35th Jubilee International Convention MIPRO under Green ICT World, Opatija, Croatia, pp. 737–742, 2012.

[19] D. Simunic and T. Dhope. Hybrid detection method for spectrum sensing in CR. In Proceedings of 35th Jubilee International Convention MIPRO under Green ICT World, Opatija, Croatia, pp. 765–770, 2012.

[20] T. Dhope and D. Simunic. On the performance of AoA estimation algorithms in CR networks. In Proceedings of the International Conference on Communication, Information and Computing Technology (ICCICT), Mumbai, India, 2012.

[21] T. Dhope, D. Simunic, and A. Kerner. Analyzing the performance of spectrum sensing algorithms for IEEE 802.11af standard in CR networks. Studies in Informatics and Control, 21(1), 93–100, 2012.

[22] T. Dhope and D. Simunic. Spectrum sensing algorithm for CR networks for dynamic spectrum access for IEEE 802.11af standard. International Journal of Research and Reviews in Wireless Sensor Networks, 2(1), 77–84, 2012.

[23] FCC Second Memorandum Opinion and Order. http://www.fcc.gov/. Accessed 2010.

[24] DD: Cognitive Access Consultation on Licence-Exempting Cognitive Devices Using Interleaved Spectrum, 2009.

[25] W. Webb. Cognitive access to the interleaved channels: Update and next steps. Presented at the IET Seminar on Cognitive Radio Communication, London, October 2010.

[26] CEPT WG SE43, Draft Report ECC 159. Technical and operational requirements for the possible operation of cognitive radio systems in the "white spaces" of the frequency band 470–790 MHz, available at http://www.ero.dk.

Biographies

Tanuja Satish Dhope (Shendkar) graduated Electronics and Telecommunication Engineering at the Cummins College of Engineering, University of Pune, in 1999. She received her Master in Elecronics Engineering from Walchand College, Sangli, Shivaji University in 2007. Currently she is pursuing her Ph.D. in wireless communication at the University of Zagreb, Croatia. Her research focus is on cognitive radio network optimization with spectrum sensing algorithms, radio channel modelling for cognitive radio, cooperative spectrum sensing, Direction of Arrival (DoA) Estimation algorithms in Cognitive Radio and in SDMA. She has published several scientific papers in journals and conference proceedings.

Dina Simunic is a full professor at the University of Zagreb, Faculty of Electrical Engineering and Computing in Zagreb, Croatia. She graduated in 1995 from the University of Technology in Graz, Austria. In 1997 she was a visiting professor in Wandel & Goltermann Research Laboratory in Germany, as well as in Motorola Inc., Florida Corporate Electromagnetics Laboratory, USA, where she worked on measurement techniques, later on applied in IEEE Standard. In 2003 she was a collaborator of USA FDA on scientific project of medical interference. Dr. Simunic is a IEEE Senior Member, and acts as a reviewer of *IEEE Transactions on Microwave Theory and Techniques*, *Biomedical Engineering and Bioelectromagnetics*, *JOSE*, and as a reviewer of many papers on various scientific conferences (e.g., IEEE on Electromagnetic Compatibility). She was a reviewer of Belgian and Dutch Government scientific projects, of the EU FP programs, as well as of COST-ICT and COST-TDP actions. She is author or co-author of approximately 100 publications in various journals and books, as well as her student text for wireless communications, entitled: *Microwave Communications Basics*. She is co-editor of the book *Towards Green ICT*, published in 2010. She is also editor-in-chief of the *Journal of Green Engineering*. Her research work comprises electromagnetic fields dosimetry, wireless communications theory and its various applications (e.g., in intelligent transport systems, body area networks, crisis management, security, green communications). She serves as Chair of the "Standards in Telecommunications" at Croatian Standardization Institute. She servers as a member of the core group of Erasmus Mundus "Mobility for Life".

Ramjee Prasad (R) is a distinguished educator and researcher in the field of wireless information and multimedia communications. Since June 1999, Professor Prasad has been with Aalborg University (Denmark), where currently he is Director of Center for Teleinfrastruktur (CTIF, www.ctif.aau.dk), and holds the chair of wireless information and multimedia communications. He is coordinator of European Commission Sixth Framework Integrated Project MAGNET (My personal Adaptive Global NET) Beyond. He was involved in the European ACTS project FRAMES (Future Radio Wideband Multiple Access Systems) as a Delft University of Technology (the Netherlands) project leader. He is a project leader of several international, industrially funded projects. He has published over 500 technical papers, contributed to several books, and has authored, coauthored, and edited over 30 books. He has supervised over 50 PhDs and 15 PhDs are at the moment working with him. He has served as a member of the advisory and program committees of several IEEE international conferences. In addition, Professor Prasad is the coordinating editor and editor-in-chief of the Springer International Journal on *Wireless Personal Communications* and a member of the editorial board of other international journals. Professor Prasad is also the founding chairman of the European Center of Excellence in Telecommunications, known as HERMES, and now he is the Honorary Chair. He has received several international awards; the latest being the "Telenor Nordic 2005 Research Prize". He is a fellow of IET, a fellow of IETE, a senior member of IEEE, a member of The Netherlands Electronics and Radio Society (NERG), and a member of IDA (Engineering Society in Denmark). Professor Prasad is advisor to several multinational companies. In November 2010, Ramjee Prasad received knighthood from the Queen of Denmark, the title conferred on him is Riddere af Dannebrog.

Online Manuscript Submission

The link for submission is: www.riverpublishers.com/journal

Authors and reviewers can easily set up an account and log in to submit or review papers.

Submission formats for manuscripts: LaTeX, Word, WordPerfect, RTF, TXT.
Submission formats for figures: EPS, TIFF, GIF, JPEG, PPT and Postscript.

LaTeX

For submission in LaTeX, River Publishers has developed a River stylefile, which can be downloaded from http://riverpublishers.com/river_publishers/authors.php

Guidelines for Manuscripts

Please use the Authors' Guidelines for the preparation of manuscripts, which can be downloaded from http://riverpublishers.com/river_publishers/authors.php

In case of difficulties while submitting or other inquiries, please get in touch with us by clicking CONTACT on the journal's site or sending an e-mail to: info@riverpublishers.com

www.ingramcontent.com/pod-product-compliance
Lightning Source LLC
LaVergne TN
LVHW012332060326

832902LV00011B/1844